THE SCIENCE OF SEWAGE

THE SCIENCE OF SEWAGE

What happens when we flush

JULIAN DOBERSKI

Published in 2024
A Pimpernel Press book for Gemini Adult Books Ltd
part of the Gemini Books Group
Based in Woodbridge and London

Marine House, Tide Mill Way,
Woodbridge, Suffolk, IP12 1AP
United Kingdom

www.geminibooks.com

ISBN 9781914902147

A CIP catalogue record for this book is
available from the British Library.

Illustrations by Sarah Pyke and John Gilkes
Designed and typeset by Danny Lyle

Printed in the UK
10 9 8 7 6 5 4 3 2 1

Contents

Chapter 1
Introduction

Poo and sewage are not pleasant or usual subjects for discussion in polite company. Yet they do generate strong emotions and anger when they appear where they are not wanted – such as on beaches or as untreated sewage pouring into rivers. This kind of waste is not something most people want to think about too often – the 'water companies' are there to deal with it. Out of sight and out of mind. But what does 'dealing' with poo actually entail? In the UK approximately 12 million tonnes (1 tonne = 1,000 kg/2,200 lb) of wastewater go into the sewers every day. This staggering volume of waste never stops flowing; there is no way to turn it off. About 2 million tonnes of it come from 45 million toilets. To put that figure into a more recognizable form, that is 2 billion litres/440 million UK gallons per day – just from toilet flushing. Because the flow is unstoppable, it is easy to see that if there is some kind of malfunction, or if the volumes exceed the capacity of the system, the sewage has to go somewhere. If not into the rivers or the sea, it will spill out of the sewer system into streets, parks and houses. Whether in rivers or in coastal waters, poo is pollution – a word that comes from the Latin *pollutus*, which refers to having been made foul. Hence the public concern.

Before we get too engaged with both the benefits and periodic inadequacies of sewage treatment in more economically developed countries, it is as well to remind ourselves that 2.4 billion or more people in the world live with no sanitation (WHO/UNICEF, 2015). This is especially true of countries in southern Asia and sub-Saharan Africa. So good (even if sometimes imperfect) sewage/wastewater treatment is a huge benefit – alongside clean drinking water – that many communities around the world can at present only dream of. While accepting the undoubted advantages of having continuous access to sewage treatment, the challenge is both to make good sewage treatment even better and to provide sewerage systems where none exist.

Having lauded the virtues of sewerage systems, it must be said that sewage doesn't normally generate much interest and sewage treatment works are typically unwanted neighbours at the back end of town. Currently, for example, there is controversy about the placement of a new sewage treatment facility in Cambridge (UK) where I live. Everybody wants their sewage 'dealt with', but not next door to them.

The main purpose of this short book is to 'look inside' the issue of sewage. That is to say, we start with a brief history of how sewage has been dealt with in the past and then move to the ecological impacts of sewage pollution in water bodies and finally discuss how modern sewage treatment works. The big question is how do you process those enormous quantities of offensive and hazardous organic waste and dirty water that are flushed into the sewers every day? What exactly happens to the poo to make it disappear? There must be a kind of (mostly reliable) biological magic involved in turning sewage into (relatively) clean water . . .

Controversies aside, the aim of sewage treatment is to receive dirty sewage water and return it, cleaned up, to rivers, lakes or the sea. This is very much a process of recycling water – that is, water abstracted from the environment is returned to the environment. It is often said, for example, that the water in the UK's River Thames is drunk several times as it travels from its source in the county of Gloucestershire, through London and out to sea. Turning drinking water into sewage, sewage into river water, river water into drinking water, drinking water into sewage and so on and so on . . . the cycle repeats.

Even though it often seems to be raining in the UK, there are potential shortages of freshwater in parts of the country where rainfall is low in relation to the needs of a large and growing population. This is true, for instance, in the region of East Anglia. As water is frequently abstracted from rivers for drinking and irrigation, effectively treated sewage water must be recycled to rivers, where it can make an essential contribution to keeping the rivers flowing. It is crucial that our rivers are kept clean, not only as a source of drinking water but also for wildlife and recreational activities such as swimming, fishing and boating. Although the seas and lakes are not usually the source of drinking water, they equally provide these other services and are also vulnerable to damage from untreated or inadequately treated sewage. This is why what happens to water and poo on their journey from toilet to treated effluent is important.

To know how the 'magic' of wastewater treatment happens does require careful reading and a bit of thought. The reward is understanding processes that are both biologically and technologically fascinating – even if you are not an engineer dealing with pipes and pumps! Part of that understanding is an appreciation of what can go wrong with wastewater treatment or in situations where the systems can't cope and sewage flows into rivers, lakes or the sea. That then leads us to consider the impact on aquatic ecosystems and how wastewater treatment systems can be enhanced to reduce the incidence of ecological harm to water resources. No wastewater treatment system can realistically eliminate all sources of pollution; the manmade world struggles to keep up with its waste. But the know-how is there to do much better if society has the will and accepts the financial price that must be paid.

The basics of 'modern' technology to treat sewer wastewater are well known and go back to the early twentieth century. Since then, there has been a continuous process of 'catch-up' as we strive to keep ahead of the rising demand for wastewater treatment and tighter environmental standards. As well as researching newer and more efficient technologies, the big challenge now is the scale of investment required to eliminate all the current inadequacies of sewage treatment infrastructure in the UK and elsewhere – to have an efficient and reliable system that cleans wastewater and eliminates the bulk of sewage-pollution incidents. This warrants a national debate irrespective of whether sewage treatment in the UK is done (mostly) by private companies or by state-owned or other public concerns. It needs more than 'easy' political headlines because the financial costs of such an investment programme will be huge and in one way or another will be reflected in sewerage charges.

This book should help you to understand the complexities of the treatment processes involved and why the current sewage infrastructure doesn't always return clean water. Although sewerage covers a range of sources from domestic to industrial, the emphasis here will be on the domestic.

The first part of the book takes a brief look at the history of sewage treatment and discusses the current structure and ownership of what are collectively called 'water services' in the UK. We then go on to consider how to assess the biological 'cleanliness' of natural water bodies and the impacts of sewage pollution. The last part of the book discusses the technical options and challenges of turning domestic wastewater into effluent that can be safely discharged into rivers, lakes or coastal waters. In other words, this last section answers the question of what goes on in a sewage treatment works.

Chapter 2
Setting the Historical Scene

An empty street on a quiet day in the UK. All appears still – but the rows of houses are using clean piped water, then flushing and draining that water as sewage into a hidden underground network of wastewater pipes. Unseen and unheard, rivers of sewage are flowing beneath urban streets. This silent and secret world is only hinted at by the manhole covers visible at intervals along roads and pavements. We can imagine the world below the manhole portal. It is an underworld giving freedom to roaming rats and teeming microorganisms exploiting the spoils of human society's waste.

So, we have clean drinking water flowing into houses while sewage makes the separate return journey. The flows of water and sewage never balance in terms of timing and volume. For example, garden watering, car washing and the bottles of mineral water or other drinks we buy see to that. And copious volumes of storm water flowing into drains and sewers can suddenly change the water balance dramatically. Gently flowing sewer pipes fill with torrents of water that sweep all before them.

As testified by past outbreaks of cholera and other diseases in Europe, any mixing of sewage with drinking water invariably leads to disastrous epidemics of disease and deaths in towns and cities. But this is not just a story from the past – it continues to haunt the lives of those people who live in unsanitary conditions around the world today. At best there may be open ditches that transport their sewage – or where the sewage merely stagnates and overflows with the next rains, when it can mix with sources of drinking and bathing waters. In Cambridge (UK), the King's Ditch – a side-channel of the River Cam – became a repository of urine, faeces and other waste in the seventeenth century. The waste was carried – probably rather sluggishly – by river water northwards through the centre of the city before rejoining the Cam, drifting past some of the city's famous university colleges along the way. Having been built on the orders of the English King Henry III in around 1267 as a defensive structure, the King's Ditch soon turned into an open sewer. It subsequently disappeared under city-centre buildings but its historical existence provides a reminder that sewage has, in the past, been an issue in Cambridge – both noxious and a likely a cause of ill health among the citizens of the city.

Throughout history, different societies have struggled to devise ways of disposing of faeces and urine 'nicely'. When human populations and densities were small, the natural environment – the fields, woods and rivers around settlements – could provide an adequate capacity for waste disposal with few environmental impacts. As societies became more populous and aggregated into villages and towns, such 'natural' disposal of faeces in back yards became increasingly unsavoury.

Some societies progressed to piped disposal of sewage. This might be thought of as a rather modern development, but various past civilizations made organizational and technological innovations that were quite remarkable at the time. For example, archaeological excavations have shown that several ancient civilizations featured pipework or channels for general drainage and/or wastewater removal. These include the use of

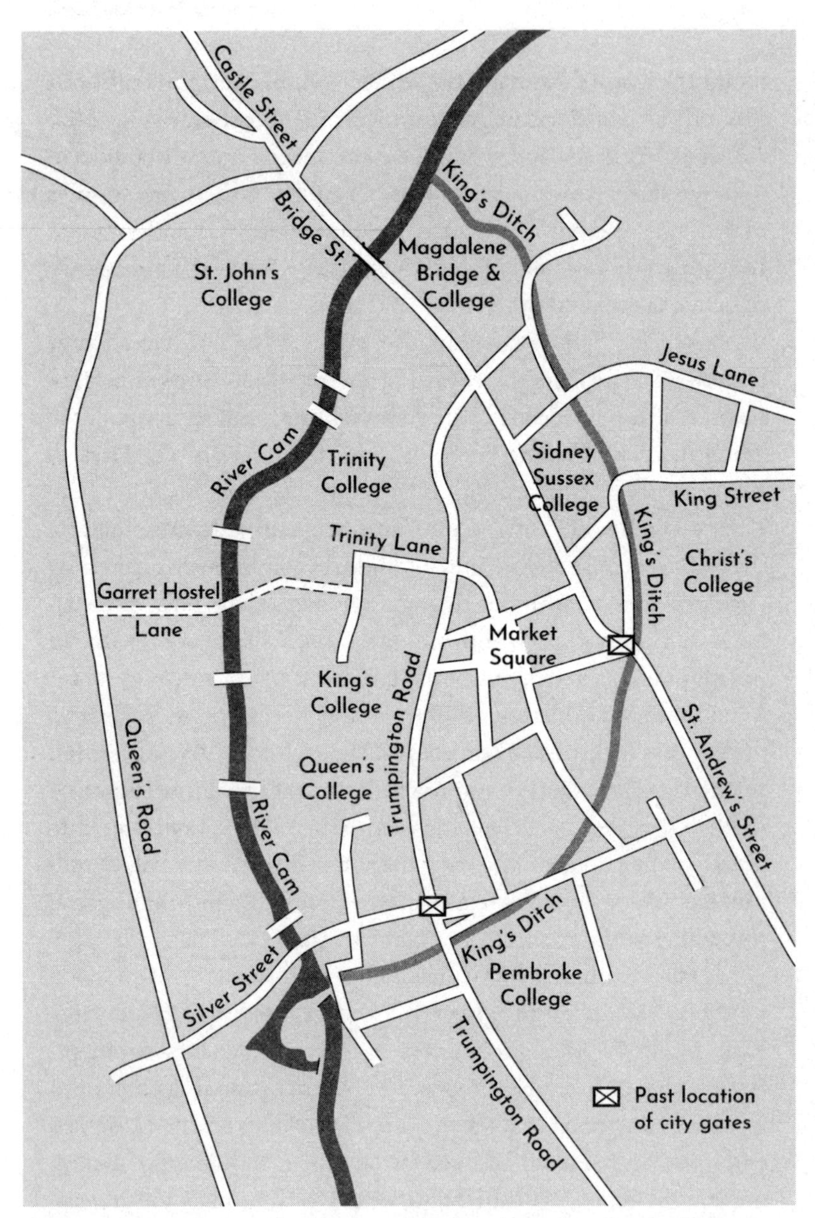

Figure 1. A map showing the location of the King's Ditch in Cambridge with the present layout of roads, which largely reflects the pattern of medieval times. University colleges have current titles.

clay sewer pipes by Mesopotamians (4000 BCE) and copper drainage pipes found in the temples of Ancient Egypt (*c.* 2400 BCE). In the Indus Valley Civilization (IVC) (largely in modern-day Pakistan) there were fully organized urban wastewater sanitation systems that drained individual houses and emptied into larger, brick-constructed street drains. The IVC was at its most advanced around 2600–1900 BCE.

Such developments in sewerage were not necessarily sustained; sometimes they were abandoned as conflict and/or societal changes led to loss of knowledge and expertise. One example of this from British and European history relates to the time when the Ancient Roman era was at its height. There is evidence of latrines channelling human waste into sewers via terracotta pipes, which were flushed with water diverted from local streams or rivers or by water drawn from beautifully engineered aqueducts. There were public and private latrines as well as bath houses feeding into the sewerage systems. Where required, lead pipes were routinely used to deliver clean water on a smaller scale – into houses, for example. However, there appeared to be no further treatment of wastewater. With adequate water flushing the sewers and emptying into larger bodies of water or elsewhere, the natural process of self-purification would decompose the waste with little ecological impact, providing the system was not overloaded.

Roman sewer systems were progressively improved, with effective public latrines known from around 200 BCE. One example in the UK is Housesteads, an Ancient Roman fort that forms part of the defensive Hadrian's Wall in northern England. Here latrines were flushed with rainwater stored in large tanks. By 100 CE, Roman sewage systems in homes and public buildings had become very efficient – such that the Roman author and natural philosopher Pliny noted that of the Romans' many accomplishments, sewers were 'the most noteworthy things of all'.

Moving forward in time, among the most elaborate water supply, bathing and sewage systems were built in the Middle East – in keeping with a stringent religious (Islamic) requirement to maintain cleanliness through washing as necessary at prayer time. Thus, during the Abbasid Caliphate (eighth–thirteenth centuries), the city of Baghdad in modern-day Iraq had 65,000 bathhouses linked to a sewerage system. Advanced water supply and wastewater systems were seen in many Islamic cities and relied on a sophisticated knowledge of hydraulics to control and direct water flows through underground aqueducts (*qanats*). The water supply also served water fountains, pools and flowing-water rills and channels in much-admired Islamic gardens.

Returning to the aftermath of Roman administration in the UK, the Romans officially left Britain in the year 383 CE, although the process of departure was somewhat protracted. Much of the infrastructure they had created then gradually deteriorated through neglect and fighting after invasions from Ireland, Scotland and Europe. The existing sewerage infrastructure was soon abandoned and little thought was given in the UK to the problem of hygienic sewage disposal for more than 1,000 years. In medieval England, townspeople used 'potties', emptying the contents out of a door or window into the street. There might be open drains or gutters in the middle of the street to collect some of this waste, which would be flushed elsewhere during periods of rainfall. In the UK the term 'kennels' was used to describe these channels. For the rich, the equivalent of Roman latrines was 'garderobes' seen in castles. These small rooms protruded from castle walls and offered a chute for directing faeces and urine into the surrounding water-filled moat. As this was typically a closed body of water, the consequences of adding copious quantities of organic waste may have been somewhat unsavoury. There was also a large public garderobe in London, flushing directly into the Thames. All in all, there were a variety of methods in 'olde England', depending on wealth and circumstances, for accommodating the disposal of urine and faeces. In Scotland in the 1700s, the tall and crowded tenements and streets of the city of Edinburgh resounded to the cry of 'Gardyloo' – a warning to passersby that a chamber pot was about to be emptied from a window, several storeys up, into the street below! The word was derived from an expression in French, meaning 'beware of the water'. The 'Nastiness Act' passed in 1749 dictated that such activities had to be restricted to the hours between 10 p.m. and 7 a.m. Suffice to say, as population crowding increased in UK towns, the environment must have become very much more foetid.

Let's move forward to a time when wastewater management starts to get really interesting. The period from around the middle of the nineteenth century to the mid-twentieth century was something of a golden age in relation to the science of sewage and advances in the engineering needed to collect, transport and dispose of sewage-contaminated water. Where do we start? The emphasis here will be on the UK, which is where the concentration of people to support the new factories of the Industrial Revolution (eighteenth and nineteenth centuries) resulted in hugely unsanitary conditions in towns that grew rapidly around the new industrial sites. A hundred people might be sharing one rudimentary toilet, with overflowing sewage potentially mixing with sources of drinking water. Serious outbreaks of waterborne diseases were common, such as the cholera epidemics of the 1830s and 1850s. Similar changes were happening in parts of Europe that were also grappling with some of the same problems linked to industrialization and the movement of people from the countryside to towns.

The first attempts to try to remedy the problem of sanitation in the UK were made in 1848, when the government issued a requirement for every newly built house to have a water closet – where water flushed away the faeces – or an ash-pit privy. The term 'privy' would normally refer to an outdoor toilet, which could be as basic as a hole in the ground or else with a pail or tub to contain the faeces and urine. The contents were typically dosed with coal ash from the house, which absorbed any liquid and moderated the smell. Of course, the privy would fill up and needed to be emptied. This was the job of the 'night-soil man', who was paid to empty privies and cart away the contents. Both in the UK and elsewhere, this night soil was typically used by farmers as a fertilizer. Although beneficial to plants, there were clearly risks associated with applying night soil to food crops – any contamination could lead to potential infection by pathogens or parasites (for example, typhoid or cholera). In addition, the importation of guano (solidified bird excrement) from South America in the nineteenth century provided a more convenient, cheap and effective form of agricultural fertilizer. The collection of night soil and its transport to farms became increasingly impractical as cities grew in size; and the rise of water-flushed sewerage eventually eliminated the need for it. Interestingly, Karl Marx noted that industrial capitalism had turned human excrement from an agricultural resource into a source of waste (Kawa et al., 2019)!

This was the state of sanitation in the UK in the early to mid-nineteenth century as the need to 'do something about sanitation and sewage' was beginning to seriously influence the thinking of some politicians, councillors and philanthropists. Scientists and engineers were thinking about the means to do it. Even artists became involved. Particularly notable was the proposal made in 1834 by the great English painter of fantastical landscapes John Martin for the creation of large interceptor sewers and walkways on both sides of the Thames in London. This was very prescient thinking, as will become apparent in Chapter 4.

Clearly there was increasing recognition that cities were becoming overwhelmed by sanitation problems and concern about the consequences of this for the health of the population. The seminal event that is generally considered to have changed the public and political mood was what became known as the 'Great Stink'. This was the River Thames at its most pungent – the hot and dry weather in June, July and August in 1858 creating an intolerable stench that permeated much of central London, reaching the hallowed halls of the Houses of Parliament. Temperatures rose above 30°C/86°F, reaching a maximum of 36°C/96.8°F, while the drought caused a marked reduction in the flow of fresh water in the river. Much of the flow in the Thames was now raw sewage. At the time, the nature of disease was largely unknown. The miasma theory of disease was the somewhat vague idea that the smells and vapours emanating from (in this case) the Thames were responsible for causing disease. Indeed, this led to the Houses of Parliament being swathed in cotton sheets soaked in calcium hypochlorite (chloride of lime) to keep the pestilence at bay. Even before the time of the Great Stink, a number of scientists and doctors had become convinced that microorganisms were responsible for disease but their work was variously rejected. It was around this time (late 1850s) that the Frenchman Louis Pasteur was strongly championing the 'germ theory of disease', which was then further developed by the German scientist Robert Koch. In the 1880s, Koch presented a list of four so-called 'Koch's postulates', which created clear scientific criteria for linking specific micro-organisms with disease – that is, establishing causality. The miasma theory of disease was laid to rest as the true nature of disease linked to sewage became clear.

If we step back a little in time to another key event, the practical and health problems of sanitation and disease in the UK were highlighted in a landmark nineteenth-century report by the social reformer Sir Edwin Chadwick. Convinced of a link between disease, ill health and poor sanitation, Chadwick and three doctors sought information and views from the working classes in cities as well as from professionals working with the poor and in city and work environments. The resulting publication, entitled *Report on the Sanitary Condition of the Labouring Population of Great Britain*, was published in 1842. The report carries the accolade of being the bestselling publication ever produced by the government Stationery Office, emphasizing the level of public interest. It led to the enactment of a Public Health Act (1848) that aimed to improve water supply and sewerage as well as rather more mundane but important matters such as street paving. Its measures were not compulsory, with the power to implement such works devolved to local areas. This resulted, as might be expected, in rather piecemeal and uneven progress.

However, increasing discussion of the issue of sewage, together with the Great Stink, provided a tipping point for sanitation reform – even the politicians were finally waking up to the problem. We will go on to discuss the consequences of the Great Stink for sanitation in the capital in chapter 4. For the moment we start elsewhere, in the city of Liverpool (UK). The developments in sanitation that were implemented in Liverpool actually pre-dated those in London. That city provided a stage for one of the first men of real vision who pioneered a radical change in sanitation thinking, followed by determined action. So, the story of innovation in UK sewerage systems really begins in Liverpool in the mid-nineteenth century.

Chapter 3
The Liverpool Story

In the early 1800s, the supply of 'clean' piped water in Liverpool was controlled by two private companies. There was no structured provision for the removal of excrement beyond night-soil collection. The flow of drinking water was frequently intermittent and generally considered unsatisfactory. With a remarkable degree of foresight, the city fathers – Liverpool Corporation – introduced two UK Acts of Parliament in 1846 and 1847 that respectively related to sewage and water supply. We will not dwell further on the water-supply issue, except to note that the Corporation's affirmative action resulted in the building of several water-supply reservoirs in the surrounding hilly/mountainous areas of England and Wales. One of those reservoirs, Lake Vyrnwy (Wales), was linked to Liverpool by a 109-kilometre/68-mile aqueduct and the problem of supply of clean water was largely solved – at least in the early days!

With regard to sewage, the 1846 Act created three linked appointments to help resolve the sanitation and health issues of the city. A medical officer of health, an inspector of nuisances and a borough engineer were appointed. This last post was filled by the hero of this story, James Newlands. He had previously been based in Edinburgh, where much of his work related to agriculture and the design of agricultural buildings. His first task on arriving in Liverpool was to have the right baseline information about the city's layout. The initial focus was therefore on producing accurate maps, which required surveyors to make around 3,000 geodetic (surveying) measurements around the city. An example of Newlands's wonderfully detailed maps is shown in Figure 2. These maps provided precise information about the layout of roads and buildings for planning a network of sewage pipes, but also included height measurements – which were essential to ensure the unimpeded gravity flow of sewage.

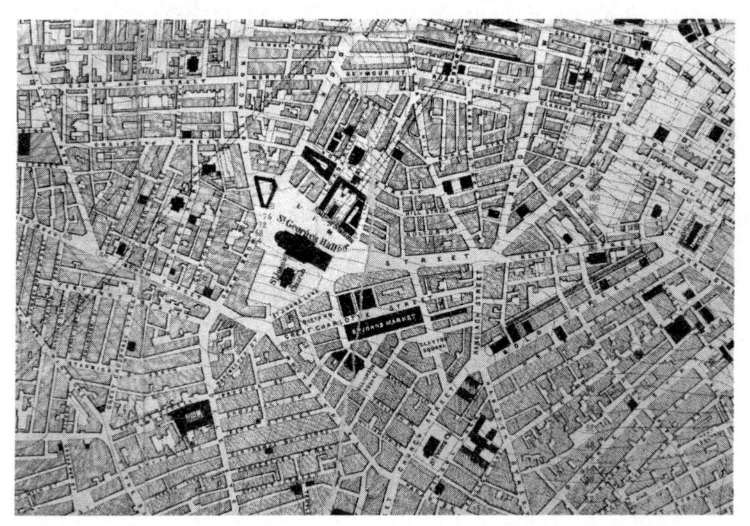

Figure 2. A fragment of James Newlands's map of Liverpool, 1848 (courtesy of Liverpool Central Library and Archives). The map includes detailed contour lines just visible in this black and white reproduction.

Having completed his groundwork, Newlands produced a report (1848) for the Corporation outlining a complete upgrading of the city's drainage and sewage system. This proposal required the replacement of privies (and more basic cesspools) with water-flushing toilets and addressed issues such as the adequate provision of public baths and wash houses. Newlands saw this very much as an opportunity to improve the living conditions of the working classes.[1]

Considering the current protracted timetables for decision-making (or lack of decision-making), it is quite remarkable that construction of Liverpool's sewerage system also began in 1848. In that way it earned the title of the first integrated sewerage system in Britain.

As a model for later UK sewerage systems, it used some important design features. The traditional sewers, which were square in cross-section, were largely superseded by ovoid or egg-shaped sewers constructed in brick. The narrow part was the base of the sewer. This was a genius design feature. It meant that sewers could accommodate large or low volumes of sewage and still function effectively in washing away solid waste. At times of low flow, 'squeezing' the water into a narrow channel ensured that fast rates of flow were maintained. Connections to the main sewer from individual houses were made using ceramic pipes, as recommended in Chadwick's 1842 report.

Newlands's 1848 plan was implemented over several years. By 1851, 27 kilometres/17 miles of the main sewer had been completed, increasing to an impressive length of 480 kilometres/300 miles by 1869. The system was well designed but

1 Not everyone was enamoured with the idea of flushing toilets. The Introduction to the published proceedings of a Congress on the Sewage of Towns (Hitchman, 1866), held in Leamington Spa (UK), noted that: 'The cause of the failure in our sanitary arrangements, and the many evils which have arisen out of them [is due to] the unnatural admixture of human excreta with large quantities of water, and to the prevailing custom of employing water as the vehicle for their removal out of houses.' The issue at the time related to the leakage of sewage water into sources of clean drinking and bathing waters and so spreading disease. Even now there remain concerns relating to the waste of valuable clean water for toilet flushing and whether better solutions are possible.

far from perfect in operation. This was due in part to a failure to implement all of Newlands's design recommendations. However, the other major problem was a lack of sufficient water to keep the system continuously flushed. Although several water-supply reservoirs had been constructed, the rapidly rising population of Liverpool meant that at times the supply was inadequate. This was eventually resolved by additional reservoir construction.

You may have noticed that little has been said about Newlands's plans for the disposal of sewage. In fact, he did note the necessity of providing a basic sewage treatment facility, but in the first instance sewage simply flowed into the estuary of the River Mersey. Liverpool has a coastal location and the area's fast tides were effective in dispersing the sewage. While the ecological impact of this may have been relatively limited at that time, in due course the city's swelling population led to gross pollution of the Mersey. Nevertheless, disposal into tidal waters became increasingly the norm for coastal towns in the UK, providing an 'easy' but environmentally dysfunctional method for sewage disposal. Liverpool did eventually acquire its first sewage treatment works, but not until the 1990s – nearly 150 years after the need for such was suggested by Newlands. A £200 million project upgraded and extended the works in 2016 to treat wastewater to a high standard. As we will see in the second half of this book, where the mechanics of wastewater treatment are described, a system referred to as sequencing batch reactors (SBRs) was used to greatly enhance the treatment of wastewater and subsequent release of treated effluent into the Mersey.

Chapter 4
The London Story

In the beginning . . .

In many ways, the preamble to the London story is not dissimilar to that of Liverpool's. Squalid living conditions for the poor, with minimal and overcrowded sanitation facilities, led to a variety of health issues and in particular disease. While Liverpool was the first city in the UK to tackle the systematic collection and disposal of sewage, cities such as Manchester and London had piecemeal sewerage systems with rivers receiving untreated sewage. In these cities, tidal cleansing was either non-existent or limited. In the case of London, we have already alluded to the high volumes of excrement that effectively turned the Thames into a stinking open sewer. It was time for radical change. The account of how this happened draws heavily on information in an excellently researched book by Stephen Halliday on *The Great Stink of London* (1999).

Change in London began in 1848 with the Metropolitan Sewers Act, which resulted in the creation of the first Metropolitan Commission of Sewers – whose members included Sir Edwin Chadwick, the author of the influential report discussed in Chapter 2. As in Liverpool, the first action point instigated by the commission was the accurate mapping of London. The commissioners also oversaw the implementation of a requirement for new houses to be constructed with a water closet or privy and a drain leading to a public sewer. In due course this requirement was extended to existing houses. Over a period of about six years, 30,000 cesspools were abolished. While the elimination of cesspools was a positive step, sewage was now being diverted into public sewers that often discharged into minor rivers (such as the Fleet) within the Thames catchment. These rivers, plus other assorted sewers, emptied directly into the Thames, generating gross pollution of the river. This brings us to the issue of tidal flow. Anyone acquainted with the Thames in central London will have noticed the substantial rise and fall of the river's water level with incoming and outgoing tides. In that sense, London can be thought of as a coastal city. Unfortunately, many of the public sewers exited the riverbank at a level that ensured they would empty only at low tide. This meant, in turn, that sewage would then flow up the river with the incoming tide rather than to the sea. So, flushing was only partially effective despite tidal water movement in the river.

Although the commission instigated useful improvements in cleansing the city of sewage, its mandate apparently didn't extend to considering how to clean up the Thames. This became a subject for debate by the second Metropolitan Commission of Sewers, which took office in 1849. This commission started to discuss the building of interceptor sewers to connect up the network of individual sewers feeding the Thames. The interceptor sewer would then aggregate the sewage at a single disposal location. At that stage, the commission stuck to the notion that the collected sewage sludge could then be spread on farmland as a soil improver and agricultural fertilizer. Hence the adoption of the term 'sewage farm' in the English language. There were two competing plans for the new sewers; and, as the commissioners could not agree on their preferred scheme, they created

Figure 3. The Fleet Ditch, seen from the Red Lion (c. mid-nineteenth century), one of several tributary rivers used as sewers and emptying into the Thames.

an open competition for schemes to solve the Thames's sewage problem. They received 137 proposals! Further arguments within the commission prevented any of the schemes receiving proper consideration. So, the second iteration of the commission was disbanded and a third established – colloquially referred to as 'the Engineers' Commission' by virtue of their membership, which included the renowned railway engineer Robert Stephenson.

Now we can bring in the name Bazalgette – the central figure in the remainder of the London story. With the establishment of a third Commission of Sewers (1849), the task of reviewing the 137 proposals was passed to Joseph Bazalgette – recently appointed as assistant surveyor to the commission. In the end, the commission decided that none of the submissions had merit so it was back to square one.

It was clear that any plan would require interceptor sewers to divert the flow of sewage from the Thames in central London and channel it for discharge into the lower reaches of the river to the east of the city. For a brief period, the third commission encouraged yet another proposal, from the in-house commission engineer and the city engineer (Frank Forster and William Heywood respectively), for interceptor sewers north and south of the Thames. Some construction work began on the north side of the river, but rising costs, poor construction, general criticism and acrimony led to the resignation of the third commission. So, yet another commission – the fourth – was constituted. They dithered about supporting Forster's plan, but further consideration of the scheme was aborted by Forster's untimely death, seemingly after a period of extreme harassment and stress.

Although his contribution is largely forgotten, Forster had laid the groundwork for the system of interceptor sewers that was finally adopted. In the meantime, his deputy, Joseph Bazalgette, was appointed to Forster's old post and (you've guessed it) a fifth commission was appointed in 1852. Interestingly, when Bazalgette was quizzed in Parliament about priorities for sewerage in London, his replies emphasized the need to improve drainage from individual houses and streets rather than worrying too much about the state of the Thames. Nevertheless, by 1854 the commission approved a modified version of the plan for interceptor sewers devised by Bazalgette and William Heywood. With everything ready for implementation, surely nothing could stop this latest scheme . . .

Sadly, it could. You will recall the social reformer Chadwick, who in 1842 had published a landmark report on sanitation in Great Britain. He had been excluded from the third Commission of Sewers onwards, but now stirred things up by persuading the government of the importance of separate sewage and rainwater collection – which was not in Bazalgette's plan. In fact, this was very forward thinking on Chadwick's part in view of modern issues that arise from mixing the two flows in one system. Nevertheless, such disagreements, along with issues of funding, led to further resignation – of the sixth commission! At this point the government, encouraged by MPs, decided to create a new Metropolitan Board of Works (1856) to take over the commission's work. This had a different emphasis in terms of its membership, with strong representation from the different London boroughs. The Metropolis Management Act (1855) that brought the new board into being envisaged the implementation of the new sewerage system under the direction of more local district boards. Was the stage finally set for progress with implementation of a comprehensive sewerage plan?

The board's first act was to reconfirm the appointment of Bazalgette as chief engineer – and to require him to produce plans for interceptor sewers at the earliest opportunity. This was accomplished in the following few months (by May 1856). Bazalgette himself commented about his plans that 'I cannot pretend to much originality', reasoning that they were largely based on several earlier plans – starting with work of Forster and others. Other commentators have been much more generous in their admiration of the project, which Bazalgette both designed and, more importantly, managed to deliver – as we shall see.

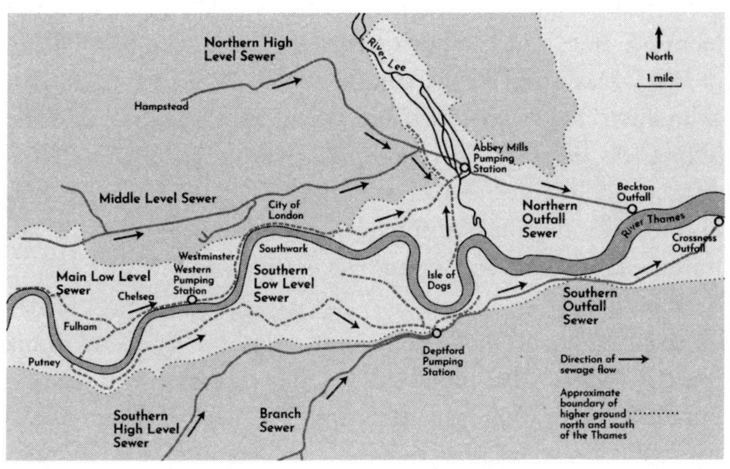

Figure 4. A map showing the location of the five main interceptor sewers in London built by Joseph Bazalgette, along with several shorter branch sewers. Major pumping stations are marked with small circles. Note the variation in level north and south of the Thames, creating a gradient for gravitational flow of sewage along those sewers that were built on higher ground. Low-level sewers to the west of London required the lifting of sewage to a higher level using the Western Pumping Station.

The plan required main interceptor sewers north (three) and south (two) of the Thames, along with a number of shorter branch sewers (Figure 4). No sewage treatment was envisaged, so the two main interceptor sewers would simply empty into the Thames close to the London Metropolitan boundary at

Barking. As far as the board was concerned, this would solve the city's sewage problem – which was then out of sight and out of mind. However, based on previous differences of view, one could not expect such a plan to be accepted without a degree of rancour. When the plans were sent by the board for government approval, there was concern that the sewer discharge points into the Thames were too close to London – bearing in mind the effect of tidal movements forcing sewage back upstream. There was further delay as alternative outfall locations were sought and costed. Fate intervened in the form of the Great Stink in the summer of 1858. As noted earlier, this caused huge consternation among MPs, the press and the public. As the *City Press* newspaper commented on 19 June 1858 on deliberations in Parliament: 'Gentility of speech is at an end – it stinks; and whoso once inhales the stink can never forget it and can count themselves lucky if he can live to remember it.'

Despite continued trenchant opposition from some quarters about the proposed location of sewage discharge points, exasperation won the day. The Metropolitan Board was given further powers to make decisions and to raise finance. At last, the project and Bazalgette really were ready for 'spades in the ground'. It had been a bruising process of decision-making when at times egos and professional rivalry seemed more important than achieving an objective of clearing the Thames in central London of sewage. Notwithstanding the reputation of the Victorian era as a period of dynamic change and progress in the UK, here there was dither, argument, indecision, politics and wasteful spending. Such issues can still impact modern infrastructure projects in the UK, which can become similarly embroiled in financial, environmental and political controversy.

Bazalgette builds

The project to provide five main interceptor sewers – as well as many smaller sewers running across central London – was an enormous undertaking. It must also have been daunting for Bazalgette to consider the endless complexities of such a build – integrating the work of thousands of workers and many contractors while still permitting the city to function. Neither was this going to be a fast build: in the end, it occupied some twenty years of construction. Inevitably the best-laid plans at the start of the project required modifications as the build progressed. The question of levels was critical because the sewerage system needed to allow as much sewage as possible to flow under gravity from central London to the outfalls. This was not an issue in parts of London where the sewers originated on higher ground, such as the high-level sewer that started in Hampstead (see Figure 4). By contrast, in Chelsea and Pimlico, where the level of existing sewers was below that of the interceptors, a pumping station was required to lift and hence facilitate the flow of sewage. The Western Pumping Station in Chelsea was built in 1875 and can be seen from trains coming into Victoria Station; it still exists and is still in use. The original design used coal-fired, steam-powered beam engines to lift sewage 6.4 m/21 ft.

There were, of course, many obstacles to overcome on the route of each interceptor sewer, be they canals, underground railways or a variety of below-ground services. In addition, the scale of the works introduced huge logistical issues related to the supply of building materials. The combined length of the interceptors was 132 kilometres/82 miles and required nearly 320 million bricks and 673,000 cubic metres/880,000 cubic yards of concrete. A vast amount of excavation was also needed, along with the disposal of the excavated spoil. All this was achieved mostly using manpower and spades. To allow for an adequate fall in level in each sewer section (a minimum of 0.61 m/2 ft per 1.6 kilometres/1 mile), each of the two main interceptor sewers required pumping stations before they reached the Thames outfall. Two major pumping stations were constructed on the east side of London, north and south of the Isle of Dogs (Figure 4 – Deptford and Abbey Mills). The most magnificent pumping station is that built in 1865 at Crossness – at the southern main interceptor sewer outfall. This pumping station is an impressive example of the Victorian tendency to go for ornate design even for structures with a seemingly mundane purpose. There was pride in the engineering involved and a desire to show it at its best. The Crossness Pumping Station was served by four giant beam engines, which continued in use until 1953. The pumping station itself is no longer operational, but the beam engines remain in place and part of the station has been restored by volunteers to its original glorious decorative state. One of the other pumping stations, at Abbey Mills, also pays homage to the achievements and pride seen in Victorian engineering, with a variety of bold decorative flourishes that enhance the building. Both Crossness and Abbey Mills are occasionally opened for public visiting – more so Crossness, where one of the original beam engines can be seen in action on open days.

The new sewerage system was officially opened in 1865 but was not fully completed until 1875. At that point it could carry 2 billion litres/440 million UK gallons of wastewater every day – with spare capacity to cope with an increasing population in London (Collinson, 2019).

One of the key features of the main sewers that contributed to their success was that they were built egg-shaped in cross-section rather than round or square. Although the design of such sewers is often attributed to Bazalgette, they were in use before his rebuild of the London sewer system. We have already noted their earlier use in Liverpool and egg-shaped ceramic sewer pipes were also used in Manchester from around 1848 onwards. However, Bazalgette does deserve credit for choosing this sewer design despite the higher construction costs of such brick-built sewers.

Of course, the missing element in the Bazalgette scheme, despite its many virtues, was the lack of any kind of wastewater processing. It took a tragic event in 1878 to bring the issue of treating raw sewage into the public eye. This concerned a collision in the Thames, just east of the two Bazalgette sewage outfalls, between a coal ship and the pleasure steamer *Princess Alice*, the latter with 600 passengers on board. The pleasure steamer sank within four minutes. One hundred and thirty passengers were rescued but a number of these people later died from disease, having ingested water grossly contaminated with London sewage.

Following a Royal Commission report, basic sewage treatment was instigated at Beckton based on chemical precipitation to remove sludge (1887–9), which could be applied to land as fertilizer but was primarily taken out in sludge vessels (locally referred to as Bovril boats) to be dumped in the Thames Estuary. A similar process was employed at Crossness after a basic treatment plant was constructed (1888–91). Surprisingly, the practice of sludge dumping in the estuary continued at both plants up to 1998.

Both Beckton and Crossness now have full sewage treatment plants constructed to treat the wastewater before its discharge to the Thames. Indeed, the works at Beckton are reputed to be the largest such plant in Europe. This is one of five major sewage treatment plants in London that have had significant investment in recent years to increase sewage treatment capacity or effluent quality. The details of the different methods of wastewater treatment currently employed in sewage treatment are discussed in the second half of this book. However, the issue of capacity in sewer systems is perennial. Cities grow and sewage volumes increase, but the sewerage systems can't cope without sufficient investment. So, the only way of dealing with the excess is to return to releasing sewage into rivers or coastal waters. In some respects, it feels like we have gone full circle, with periodic or even frequent pollution events in rivers caused by excessive sewage volumes in combined sewerage systems. This has led to justifiable concerns and complaints from a variety of pressure groups and members of the public who have an interest in the health of rivers and coastal waters. Once again, headlines akin to 'Something needs to be done' are splashed across newspapers and other news outlets. In the next chapter, we look at two major projects designed to improve river-water quality in London and Paris.

Chapter 5
Sewage Pollution in London and Paris: The Thames and the Seine

We have already noted that minimizing the impacts of sewage on rivers flowing through cities is a never-ending task. Not only do most cities grow – as does the volume of sewage – but public expectations of water quality also rise, together with a general tendency for lawmakers to raise environmental standards. At the same time, in most older cities there is likely to be an old sewerage infrastructure to work with. Despite this – and even though change can seem glacially slow and often imperfect – overall, there have been major improvements in the quality of effluent discharged from sewage treatment plants in UK cities.

So, what about the Thames and the Seine? Both rivers have continuously been in a cyclical process of 'catch-up' as they have grown in population. Deterioration of water quality in both rivers due to periodic but steadily increasing releases of excess wastewater from the sewerage system is then countered by major investment, which significantly reduces environmental degradation and improves water quality – at least for a while. We will look at two examples of current investments related to these two rivers that should significantly improve water quality for some years.

The River Thames

Starting with the Thames, we can ask: what is the state of the water in that river? A 'laboratory' approach to water quality is to measure a variety of chemical parameters to give a direct indication of the extent of pollution in the river. Ideally this would be based on continuous chemical measurement rather than spot measurements, to cope with the vagaries of water flow and wastewater releases. The other way of assessing water quality is to look at the living organisms in the river – the biota. The survival of living organisms depends on the effects of a host of major and minor factors – from major one-off pollution incidents to the insidious effects of 'minor' pollutants whose long-term effects may be largely unknown.

The biological indicators of water quality that have been most widely studied in the Thames, and that also resonate with the public, are the fish. Before about 1800, the Thames was relatively clean and could support large fish populations (Richardson and Soloviev, 2021). The nature of the species present was affected by one factor that continues to this day: the effect of salinity. Saltwater flows up the river from the estuary with the incoming tide. It currently reaches west of London to the town of Teddington, where a weir prevents further ingress. So, in the past, fish populations have always included species that tolerate low-salinity (brackish) water, as well as fish species that alternate in their life cycle between sea and freshwater (diadromous species).

It was this mixed-species population of fish that began to be severely affected by wastewater pollution – particularly faecal organic matter – from the 1800s onwards. Later, several bankside power stations releasing cooling water raised the river-water temperature, reducing the oxygen-carrying capacity of the warmer water. The volume of raw sewage releases has since much reduced – it's now only about 12 per cent of historical volumes in the mid-nineteenth century (Richardson and Soloviev, 2021) – and the power stations have gone. However, there are additional inputs of pollutants from farming operations adjacent to both the Thames and its tributaries. In addition, sewage effluent now contains many novel molecules that may not be removed by wastewater treatment (for example, medications such as antibiotics and other pharmaceuticals, nanoplastics and microplastics). However, it is the periodic release of organic waste from sewerage systems when at full capacity that continues to have a major impact on the Thames and can still create oxygen-deficient zones.

So, what has happened to the fish population in more recent times in the still challenging aquatic environment of the Thames? Well, fish have returned to the Thames in central London since it was labelled 'biologically dead' in 1957 by the London Natural History Museum. However, fish continue to 'run the gauntlet' past periodic sewage effluent overflows as they swim along the river. Around 115 fish species are now listed for the Thames. This sounds an impressive number, but many of these are rare. In addition, the damaged river ecosystem has helped a variety of non-native (exotic) species to add to the fish community. So, the reality is that this is not a 'natural' fish community – instead, the biomass is dominated by a small number of more pollution-tolerant fish species, such as freshwater sticklebacks and roach, and marine species like whiting and herring (Richardson and Soloviev, 2021).

An article in *The Independent* newspaper recently asked the question: 'How did the Thames become one of the world's cleanest city rivers?' (Edmonds-Brown, 2022). It is undoubtably true that the river is much cleaner than it was, but releases of untreated wastewater mixed with surface drainage (combined sewerage) continue at times of rainfall when the sewer capacity of the system is overwhelmed. Past contaminants persist in riverbed sediment, including heavy metals and a range of other pollutants. With the current capacity of the London sewerage system as it is (in 2024), even light rain can cause sewage to overflow. This problem can be solved only by providing enough capacity to retain large volumes of storm water and sewage until it can gradually be transferred to a wastewater treatment plant.

The answer to the problem in London has been to plan a single major sewer tunnel, about 25 kilometres/15.5 miles long and 7.2 m/23.5 ft wide, running east–west, broadly following the path of the Thames. At the time of writing, this is being constructed and is referred to in publicity as a 'super-sewer' – or more officially as the Thames Tideway Tunnel. It is currently scheduled to be completed in 2025 at a cost of over £5 billion (or over $6 billion). How does it answer London's sewage problem? Well, in a manner not dissimilar to the Bazalgette scheme, it is an interceptor sewer, but one built to a maximum depth below ground of 70 m/230 ft. It doesn't replace Bazalgette's interceptor sewers; rather it drains thirty-four major sewer overflows in the city. Eventually the sewage it collects flows into the Stratford to East Ham (Lee) tunnel constructed in 2016, which conveys the sewage to the Beckton Sewage Treatment Works. Not only does the Tideway Tunnel have an immediate effect on diverting most (but not all) sewage overflows from the Thames, but its sheer size also provides temporary storage capacity for exceptional levels of rainfall. It is argued that this storage capacity will serve the increasing population in London for about 100 years. And if the improved water quality lives up to expectations, it will allow Londoners to reconnect with their river – providing leisure opportunities and pleasure from the return of nature to a much cleaner Thames.

This is one way of solving the sewage capacity problem in a city. For an alternative way we can look at Paris.

The Seine

Rather like the River Thames in London, the Seine in Paris has long been plagued by poor water quality. One difference between these two rivers is that the Seine is not tidal – sewage is swept downstream rather than being shunted back and forth by the tides. That aside, the issue of storm overflow of sewage is common to both cities. There are forty-four separate outflows in the case of the Seine. Spurred on by the award of the 2024 summer Olympic Games, Paris has taken on the task of a serious clean-up of the Seine. The ambition is focused on a specific end-point – to allow safe urban swimming in clean river water and a safe location for Olympic sporting events. As in London, the issue is one of retaining mixed storm drainage and sewage at times of high rainfall before it is then gradually pumped away to treatment works. In Paris, the solution involves the building of a massive underground storage tank and a budget of €1.4 billion (£1.2 billion/$1.5 million). The tank is located under a central Paris public park on the Left Bank of the Seine, behind Austerlitz train station. It is designed to accommodate up to 46,000 cubic metres (over 10 million UK gallons) of mixed rainwater and sewage – broadly equivalent to thirty Olympic-sized swimming pools. Impressive as it is, it seems that it will not entirely solve the overflow problem. Water quality will be much improved, but heavy rainfall may continue to cause periodic issues of sewage contamination.

So, there we have it – a tale of two cities and a pointer to the kind of massive investments needed to increase capacity in legacy sewerage systems from the nineteenth century. Of course, we have only referred to two capital cities. Many countries, cities and urban conurbations have similar legacy issues. The question of who should run water and sewerage enterprises – and so be responsible for financing the massive cost of sewerage infrastructure improvements – is raised in the next chapter.

Chapter 6
Water Services: Who Runs Them, Who Pays for Them – and How?

Ownership of sewerage systems is currently under the spotlight in the UK because of perceived failings on sewage pollution. There is no single model of water services ownership that applies to the whole of the UK. Instead, there are four main options. The first is individual private ownership of sewage disposal – especially in remote areas where basic sewage treatment involves the installation of a septic tank (or other methods) not connected to the sewerage system. This option doesn't bear directly on the present discussion. The other three main options are public ownership, public-private ownership partnerships or fully private ownership of water supply and sewerage – collectively referred to as water services. Complete private ownership of the water services infrastructure is a particular feature of the industry in England, a model that (almost) no other countries have chosen to follow.

So, the question of water services ownership has become rather contentious in England: should it be public or private – or something in between? This chapter won't attempt to answer that question, but it will briefly review some of the issues that relate to it. These have also been well rehearsed in various articles in quality newspapers such as *The Guardian* and in television and other media reports.

The current heightened level of public interest in this issue has been triggered by news of numerous releases of water contaminated with sewage into some UK rivers, lakes and coastal waters. This is not an acceptable method of sewage disposal, but occurs when there is excessive rainfall draining through a combined sewer system that cannot contain the volume of effluent – these are known as combined sewer overflows (CSOs). Concern about the effects of such spills on both health and the environment has galvanized protests by pressure groups such as surfers, swimmers and anglers, among others. They accuse water services companies of unjustified releases of sewage, even at times when rainfall is not an issue, and of not investing enough to solve the CSO problem.

There is little argument about the two main causes of the sewage overflow. The first of these is the unfortunate design of earlier sewerage systems that combined sewage with relatively clean rainwater. There are large areas of the UK where combined sewers are the norm and there is insufficient storage capacity in the system for temporary retention of excessive flows when this is needed. This is, of course, exactly what was discussed in the previous chapter in relation to London and Paris. Those two examples teach us two things: that it is possible to do something about the sewage-overflow problem but also that enormous sums of money are required to fund such improvements. A UK Parliamentary report noted that the complete separation of the 100,000 kilometres/62,137 miles of combined sewers in the UK would cost £350–£600 billion – an unrealistic level of expenditure. Any such programme for large-scale replacement of combined sewers would also be massively disruptive in towns and cities (Tudor, 2022).

In reality, no one likes to pay more for water services if someone else can shoulder the financial pain. But someone must pay and ultimately that generally means the consumers of water and sewerage services. The riposte from consumers is typically, 'Why doesn't the government pay?' or 'Why don't the private companies pay instead of giving money to shareholders?' Yet, if the industry is in public hands (which it is not in England, but often is elsewhere – including Wales and Scotland),[2] governments can either spend money from taxes to invest in water and sewerage infrastructure (which the consumer pays for through higher taxes or water charges) or borrow money (which consumers repay through their bills, along with the interest payments for borrowed money). Of course, governments may choose to avoid making such investments if, for example, there are economic concerns about the level of public spending, borrowing and government debt. Government involvement doesn't guarantee investment – politics can get in the way.

What about the private companies that own the English water services industry? They are owned by private investors – an assortment of banks, private equity companies and sovereign wealth funds whose aim, as in any free-market economic system, is to make a profit and disburse those profits to their shareholders. That is the simple explanation of what they do – in reality, there are varying degrees of what is referred to as financial engineering that can make the finances of such companies seem rather opaque. One example that has been cited in the media is the payment of dividends to shareholders while adding those payments to the debt burden of the company. What is clear is that companies that received a dowry of zero debt at the time of privatization by the UK government must now service huge levels of debt – estimated to be around £54 billion/$67

2 In Wales there is a kind of hybrid system – a private water services company with no shareholders. It is owned by the Welsh government and any profits are reinvested. This company has also been accused of illegally discharging sewage.

billion (Guardian Editorial, 2023; Leach et al., 2023).

Some of those debts arise when these private companies borrow money for major infrastructure investments (such as the Tideway Tunnel described in Chapter 5). That funding can come from various banks or shareholders or elsewhere. The debt also needs to be paid back, along with interest payments on the borrowed money. Ultimately it is the consumer who pays for this as well.

All of this complex financial activity should be scrutinized and approved in the UK by the government body Ofwat – the Water Services Regulation Authority. It regulates the financial dealings, water bills and corporate structure of water services companies as well as agreeing performance levels and investment plans. If that level of investment has turned out to be inadequate, then there are questions to be asked about why Ofwat has not been more ambitious in directing companies to invest to improve their environmental performance. It may be that they have sought to balance the need for investment against large increases in water services bills that would have resulted from more ambitious expenditure plans. So here again, the needs of politics may trump environmental improvements.

Although this is a rather superficial review of the options for ownership of the water services sector, it should be clear that, while the current system can fail in maintaining a high standard of water quality in rivers and coastal waters, a fair degree of blame for this can be apportioned to the UK government and Ofwat – as well as to water services companies. Water services companies operate within a monopoly system (in each region) created and controlled by the government. Collectively all parties need to do better – but that will inevitably lead to higher water services bills. There has been talk of renationalizing water services. It has been proposed that this could be done without compensation in view of the debt liabilities that the companies carry – but then those liabilities would need to be transferred to the government. It doesn't seem likely that there will be wholesale change in the system any time soon. Even if water services were renationalized, there would be no assurance of rapid investment by government towards a quantum change in environmental performance. The debate continues (Laville, 2024).

Chapter 7

Measuring the Impacts of Sewage and Treatment Effluent on Water Bodies

Self-purification of rivers and lakes

No 'clean' river or lake is ever free from some organic 'pollution'. Both rivers and lakes cope with varying natural inputs of organic matter, ranging from bird and fish faeces to the annual deluge of deciduous tree leaves in autumn. This is all part of the natural ecology of a river or lake. Both are ecological systems with a range of plant, animal and microbial components. We can think of such ecosystems as a large food web with many food-chain links representing the feeding relationships of a host of aquatic organisms. Aquatic plants and algae contribute majorly to capturing light energy to power the ecosystem through their growth. This productivity is supplemented by inputs of organic matter from terrestrial sources outside the river or lake. Hence, organic matter in rivers naturally comes from two sources. There will be dead plant, animal and microbial organic matter generated internally in the water body (autochthonous material) and external inputs such as dead leaves shed from river/lakeside vegetation or blown into the water (allochthonous material). What then follows is a process of decomposition of this organic matter, which will be fast for some nutrient-rich materials, but slower for organic molecules that are hard to decompose. Fallen leaves and aquatic plants go through an initial process of decomposition of 'easy' parts – the contents of plant cells, which may contain a nutrient-rich mix of sugars and proteins and other molecules. But the tougher parts of the leaves, rich in cellulose, may take months or longer to decompose – a process that isn't helped by the cold winter weather following leaf fall.

Of course, all this decomposing plant material is a source of energy and nutrients. It typically supports many species that form the decomposer community – microorganisms, invertebrate detritivores, microbivores, fish and invertebrate predators of the decomposers. A range of invertebrate consumers shred and feed on the dead plant material. In addition, herbivores (vertebrate and invertebrate) feed on living aquatic plants and algae. So, both living and dead organic material drives the functioning of the aquatic ecosystems. It is a resource that provides energy to power the cell and tissue biochemistry (metabolism) of the water-based animals and much of the microbial aquatic community. In addition, organic material provides the nutrients that are the 'bricks' that build the bodies of animal, plant and microbial organisms. So, what kinds of organisms are found in freshwater communities? That will depend on the type of water body – still or flowing – and a range of other physical and chemical factors. Species living in the silty bed of a lake or pond will be different to those in the gravelly bottom of a fast-flowing river or stream. These may include a selection from birds (of aquatic habitats), fish, frogs, toads, newts and a variety of mature and immature insects, aquatic crustaceans, various worms and single-celled protoctists (protozoa) such as the familiar *Amoeba*. More cryptic, harder to identify and enumerate are the microorganisms, including fungi, algae, photosynthetic and non-photosynthetic bacteria and viruses. There may well be a visible layer of 'slime' on the surface of plants, stones or gravel in the water. A German term – *aufwuchs* – is used for this material, which is also known by the scientific term periphyton. This is a varied mix of different bacteria, algae, invertebrates, debris and complex materials (polymers) secreted by the microbial cells – probably the closest you can easily get to spotting microbial growth without a microscope. This layer of slime can also be labelled as a biofilm – a term we will encounter later in the book in the context of sewage treatment.

The decomposer community represents the machinery of decay and nutrient recycling in the lake or river and is responsible for 'self-purification'. This term is applied to the natural process of keeping rivers and lakes 'in balance' in relation to natural organic inputs and nutrients. But self-purification has a finite capacity – too much input of organic matter/waste and things start to go awry. The river or lake can no longer cleanse itself.

Sewage in coastal waters

As described in the stories of sewage in the cities of Liverpool and London, coastal waters have frequently been the recipients of large quantities of unprocessed or barely processed sewage. These have been pumped directly into coastal marine waters via offshore outfall pipes, or into estuarine waters where freshwater combines with seawater to result in brackish water of varying salinity. The extent of coastal water pollution by sewage has generally reduced in recent years, but it continues to significantly impact coastal water quality. Self-purification may occur effectively through rapid dilution where there are strong currents and tidal flow, but there are still issues of local contamination of bathing waters by pathogenic bacteria and viruses (and sewage waste) if dispersal is insufficient. The dumping of sewage sludge in the Thames Estuary was a prime example of sewage overwhelming the process of natural self-purification, which led in turn to the creation of dead anoxic zones – where oxygen is absent.

Measuring the 'cleanliness' and biological 'health' of water bodies receiving sewage

What is the baseline for a healthy water body in terms of its 'cleanliness' and biological 'health'?

This is not an easy question to answer as all water bodies will have a variety of natural chemical, physical and biological characteristics. But, for a start, there are several chemical and physical parameters that could indicate that something 'unnatural' is going on. Examples of such characteristics include measures of water acidity/alkalinity, a variety of dissolved nutrients (for example, nitrate ions) and turbidity. The latter relates to the presence of particles in suspension, clouding the water.

But what about the biological health of the river or lake? The living organisms present act as silent sentinels ('canaries in the coal mine') of the physical, chemical and biological water quality. Individual species or whole groups of related species (taxa) may appear or disappear in response to changing conditions. Hence the notion that one can monitor the state of the river by scoring the living organisms that are present in it. As already suggested, the range of species is potentially very wide – from plants to fish to invertebrates to microorganisms. They can't all be monitored, so which to use?

Microbial indicators

The obvious response is to start with the bacteria. If you want to know whether there has been a recent discharge of sewage into a river (or other water body) or whether the water is microbiologically clean, then it follows that looking for 'sewage bacteria' makes sense. This means looking for bacteria associated with the gut of humans or other animals like cattle. Although there are a number of candidate bacteria, the key species belong to a group (family) called the Enterobacteriaceae – of which the best known, and the cause of sickness in humans, is *Escherichia coli* (typically referred to as *E. coli*). At its simplest, you can think of taking a water sample and putting a standard volume of the water on a nutritive agar medium (jelly-like) to grow up and count the bacterial colonies. In practice, the bacteria are typically filtered out on to a membrane filter and then incubated on a selective agar medium at a high temperature of 44°C/111.2°F, which is selective for the thermo-tolerant *E. coli*. The term selective refers to the fact that only the bacteria of interest will grow on a particular agar medium or at a particular temperature. This avoids the culture being swamped by the many different bacteria that will be present in the water sample. Counting the colonies not only tells you that *E. coli* is present, but also provides a quantitative index of the degree of bacterial pollution. The more colonies, the more *E. coli* pollution. The cultivation of the bacterial colonies to make them countable takes a day or two and unsurprisingly there are now several competing enumeration technologies for faster results. Either way, these methods provide a measure of *E. coli* pollution. Other groups of bacteria may also be measured. In the case of the UK, the microbiological standard for 'natural' bathing waters includes total counts of intestinal enterococci, such as *Enterococcus faecalis*, a cause of urinary tract infections and meningitis, among other infections.

What do the colony counts tell us? *E. coli* is not in itself a particularly significant measure for the ecology of the river. As we will see, other components of sewage effluent cause a lot more problems. However, bacterial counts can indicate that the water body is polluted by sewage and, of course, that this represents a potential hazard for human health – never more so than if people go bathing or swimming in it. In the UK, the Environment Agency sets standards for what levels are acceptable for 'bathing waters' (see Table 1).

Classification	Thresholds
Coastal bathing waters	
Excellent	EC: ≤250 CFU/100 ml; IE: ≤100 CFU/100 ml (95 per cent)*
Good	EC: ≤500 CFU/100 ml; IE: ≤200 CFU/100 ml (95 per cent)*
Sufficient	EC: ≤500 CFU/100 ml; IE: ≤185 CFU/100 ml (90 per cent)*
Poor	means that the values are worse than the sufficient
Inland bathing waters	
Excellent	EC: ≤500 CFU/100 ml; IE: ≤200 CFU/100 ml (95 per cent)*
Good	EC: ≤1,000 CFU/100 ml; IE: ≤400 CFU/100 ml (95 per cent)*
Sufficient	EC: ≤900 CFU/100 ml; IE: ≤330 CFU/100 ml (90 per cent)*
Poor	means that the values are worse than the sufficient

Table 1. The UK microbiological standards for grading bathing waters. Note that bathing is advised against only when the water is classed as 'Poor'.

Key: EC = *Escherichia coli*; IE = Intestinal enterococci; CFU = colony-forming units.

*Numbers in brackets relate to percentile values – either a 95 per cent or 90 per cent percentile depending on the classification. This is a statistical measure derived from multiple samples, which explains what look like inconsistencies between standards for 'Good' and 'Sufficient' (95 per cent is harder to achieve than 90 per cent).

Invertebrate biotic indicators of water quality

One of the problems with monitoring rivers and lakes for water pollution with laboratory-based chemical measures is that in most cases the water sampling is sporadic and with long intervals between samplings. Although it is technically possible to have continuously recording instrumentation measuring some types of pollution, this is seldom feasible when many different water bodies must be monitored. Ideally, the monitoring method should be one that integrates the ever-fluctuating conditions of the water body to give a realistic picture of 'cleanliness' between sampling intervals. For example, periodic sampling can miss short-term pollution incidents that may be infrequent but nonetheless have lasting impact on river fauna and flora. Such sampling is also unlikely to pick up the effects of low concentrations of a plethora of chemical pollutants that can now find their way to water bodies. For this reason, rather than relying solely on chemical and physical measurements, the emphasis has switched to monitoring organisms, especially larger aquatic invertebrates – referred to as macroinvertebrates. These comprise a wide range of species that can respond in different ways to pollution and other environmental challenges. The most successful monitoring method based on macroinvertebrates (referred to as a biotic index) is the biological monitoring working party (BMWP), developed in its original form in about 1980. The BMWP method works on the basis of identifying macroinvertebrate families – or in some cases genera – found in the water body. It would be too difficult and too time consuming to monitor a large number of rivers and samples at the species level. Originating in the UK, the BMWP method has now been locally adapted and applied in a many different countries from temperate to tropical. Although it has been modified for its current use in the UK, it is described here in its original form for simplicity.[3] More information on revised versions of the BMWP biotic index system can be found in Paisley et al. (2014).

Kingdom: Animalia (animals)
Phylum: Arthropoda (arthropods)
Class: Insecta (insects)
Order: Ephemeroptera (mayflies)
Family: Baetidae (a mayfly family)
Genus: *Baetidae* (one subgroup within the mayfly family)
Species: *rhodani* (one of the species in genus *Baetis*)

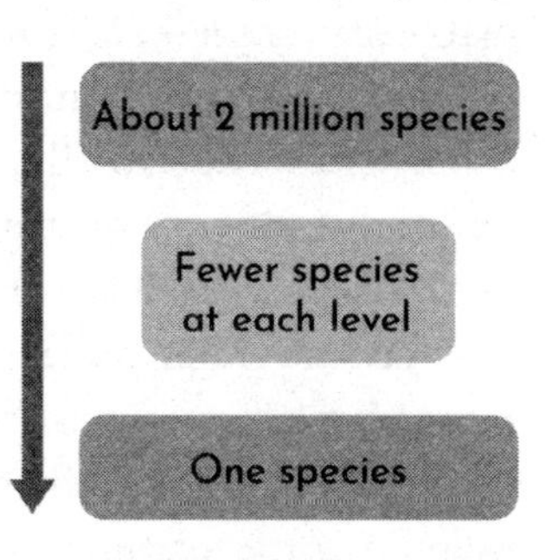

Table 2. An example of how a common river macroinvertebrate is classified biologically to species level: the large dark olive mayfly – *Baetis rhodani*.

Although lakes and rivers can seem far from biodiverse when looked at from the surface, the use of a pond net quickly reveals a variety of large and small invertebrate types (taxa). Each of these occupy a particular part of the lake or river ecosystem where conditions suit their physiological and behavioural needs – their ecological niche. These requirements vary between species, but different families can also show broad differences in tolerance to organic pollution. A particular group of invertebrates will remain

3 BMWP has been replaced for current biological river assessment in the UK by another metric, Walley Hawkes Paisley Trigg (WHPT). This is based on similar principles to BMWP but records a larger range of taxa (families) and includes abundance categories. The metrics WHPT generates are in line with those from BMWP scoring (Macadam, 2021). It has been argued that invertebrate identification at family level is insensitive to differences in pollution tolerance within families. Detailed surveillance of the water quality of individual rivers can benefit from greater taxonomic precision. So biotic indices like BMWP and WHPT are a compromise - balancing practicality when sampling many rivers at numerous sites against taxonomic exactness and the additional useful information that might provide.

in the water body only if all of the conditions that affect them stay within the boundaries of what that group can tolerate. For example, if the level of oxygen in the water falls below the threshold for a group of organisms, then they die out. So, it is a natural thought experiment to imagine that monitoring such species will reflect instances of adverse conditions in the water. This could be due to pollution but could also be caused by natural events such as unusually high water temperatures or low water-flow rates, impacting the oxygen content of water and causing mortality.

Of course, you need a good biological understanding of how different invertebrate families respond to varying conditions to interpret invertebrate sampling data. For some families it is necessary to further separate organisms into genera. And you need to integrate the results of monitoring across a wide range of invertebrate groups to give you an overall picture of the state of the water. For example, if you know of several different invertebrate types that are sensitive to oxygen levels in the water, you might look for a response from all of them as a pointer to low oxygen levels in water. This would be indicated by a decline in abundance of those families.

Sensitivity to environmental conditions is the basis of the BMWP system. The first point to note is that the method was initially designed for a particular type of water body – that is, relatively shallow rivers, which can be sampled fairly easily by wading in. Sampling takes the form of a three-minute kick sample: walking into the river, agitating the substrate and aquatic plants with a foot and allowing the current to carry the dislodged invertebrates into a net (Figure 5). Where there are both still areas of flow (pools) and faster-flowing areas (riffles), both should be sampled.

After tipping out the collected sample from the net, the invertebrates need to be identified – mostly to family level,

Figure 5. Kick sampling for aquatic invertebrates. 'Kicking' the riverbed disturbs bottom-living invertebrates, which are then carried by water flow into the net.

but sometimes to genus. These are quite broad taxonomic categories, but identifying the invertebrates *in situ* still requires a degree of expertise from experienced field staff. Otherwise, samples need to be returned to the laboratory for identification, which is a more unwieldy process. A selection of the types of invertebrates and scores that contribute to a BMWP assessment is shown in Figure 6, while the BMWP scores allocated to different taxonomic groups are listed in Table 3. As noted earlier, different organisms have different ecological tolerances. Some have very narrow tolerance ranges for particular ecological conditions and so even a small change has a big impact. These are the invertebrate families which get a high BMWP score – they are very sensitive biotic-indicator species of clean water. Overall, sensitivity to oxygen levels in the water is the factor that especially differentiates invertebrate groups in scoring low or high in BMWP. Those invertebrate families that require high concentrations of dissolved oxygen get a high BMWP

Mayfly nymph - BMWP score: 10 or 7 or 4

Stonefly nymph - BMWP score: 10

Dragonfly nymph - BMWP score: 8

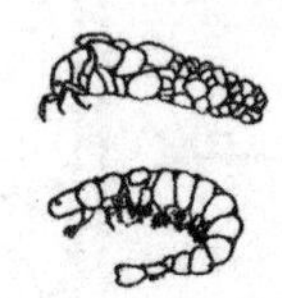

Caddis-fly larvae - BMWP score: cased (top) 7; caseless (bottom) 7 or 5

Freshwater shrimp - BMWP score: 6

Freshwater limpet - BMWP score: 6

Water bugs - BMWP score: 5

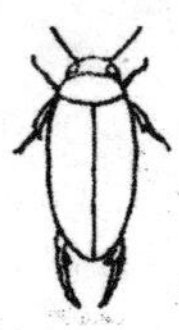

Water beetles - BMWP score: 5

Fly larvae (Simuliidae) - BMWP score: 5

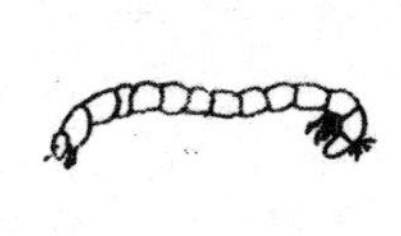

Fly larvae (Chironomidae) - BMWP score: 2

Flatworm - BMWP score: 5

Alder fly - BMWP score: 4

Leeches - BMWP score: 3

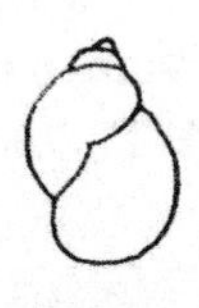

Snails and bivalves - BMWP score: 3

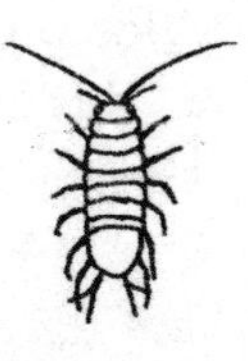

Water hoglouse - BMWP score: 3

Water mite - BMWP score: 3

True worms - BMWP score: 1

Figure 6. A selection of macroinvertebrate groups that are used in scoring biological water quality using the BMWP biotic index, as listed in Table 3. Sizes are not to scale and will vary from just 1–2 mm (water mites) to several centimetres (for example, worms and leeches). Where different families within the same order vary in their sensitivity to organic pollution, there is more than one BMWP score.

Groups (families)	Score
Mayfly nymphs (e.g. Ephemeridae, Ecdyonuridae); stonefly nymphs (all families)	10
Damselfly and dragonfly (all families); freshwater crayfish (Astacidae)	8
Mayfly nymphs (Caenidae only); cased caddis larvae (all families); caseless caddis larvae (e.g. Rhyacophilidae)	7
Freshwater shrimp (Gammaridae); freshwater limpet (Ancylidae)	6
Water bugs (all families); water beetles (all families); caseless caddis larvae (Hydropsychidae only); fly larvae (Simuliidae, Tipulidae); flatworm (all families)	5
Mayfly nymphs (Baetidae only); alderfly larvae (Sialidae)	4
Snails (e.g. Lymnaedae, Planorbidae, Physidae); freshwater bivalves (Sphaeriidae); leeches (all families); water hog louse (Asellidae); water mites (all families)	3
Fly larvae (Chironomidae only)	2
True worms (all families)	1
Fly larvae (e.g. Culicidae, Tabanidae, Chaoboridae)	no score

Table 3. BMWP scores for the range of taxa that contribute to a BMWP assessment. Although given scores are often the same for all invertebrates belonging to the same invertebrate order (for example, all families of dragonflies and damselflies in the order Odonata), in some cases different families in the same order are allocated different scores because of different tolerances (for example, Mayflies in the order Ephemeroptera). Note that this list does not include all possible invertebrate types that may be found. Not all invertebrates are allocated a BMWP score.

score – because clean rivers are generally well oxygenated. A 'clean' lake will also have adequate dissolved oxygen, although it will generally be at lower concentration and replenished more slowly in still water and a different suite of invertebrates will therefore be present. Oxygen levels are depressed by organic sewage pollution, which will therefore reduce or eliminate the high-scoring BMWP species.

How is this scoring system used in practice? The first point to stress is that when using the BMWP biotic index, invertebrate families are recorded solely on the basis of presence or absence. So, for example, finding just one or a few stonefly nymphs gives a score of 10 and so on for the other families.

BMWP score	**Biological quality**	**ASPT**	**Water quality**
Over 130	A. Very good biological quality	Over 7	Very good (natural)
81-130	B. Good biological quality	6.0-6.9	Good
51-80	C. Fair biological quality	5.0-5.9	Fair
11-50	D. Poor biological quality	4.0-4.9	Poor
0-10	E. Very poor biological quality	3.9 or less	Very poor

Table 4. Interpreting the scores from BMWP biological water-quality monitoring.

Key: BMWP = biological monitoring working party;
ASPT = average score per taxon.

After all of the catch has been checked, there is then a simple process of adding up the individual values for each scoring group (taxon) recorded to give a total BMWP score – as shown in Table 4. But there is an issue that can affect the interpretation of water quality using BMWP scores. You can have a situation in which a relatively large number of low-scoring groups (taxa) are present, which then gives a misleadingly high BMWP score. The biological quality then appears better than is justified by the invertebrate groups (taxa) recorded. This can be checked using a second metric, shown in Table 4 – the ASPT or average score per taxon. Again, a simple calculation: the total BMWP score is divided by the number of scoring groups (taxa) found. If all is well, the BMWP and ASPT scores will coincide in the water-quality assessment. If not, then interpretation of scores needs to be more nuanced.[4]

Such scores also need to be judged based on the physical/ chemical character of the river in question. In particular, factors like water pH can influence scores, irrespective of any pollution issues. Thus, a river with acidic water pH (<pH7)[5] will tend to produce lower BMWP scores than more alkaline water (>pH7). Adjustment can be made for this by using BMWP scores in a different way. Based on many surveys of different types of river with clean water, it is possible to predict what invertebrate score a perfectly unpolluted river of that type would achieve. The BMWP scores can then be compared against a predicted BMWP score for that river in a clean state. This provides a 'cleanliness' assessment of the river that is based on invertebrate surveys and accounts for the variety of types of river with different climatic, geological and geographical locations (see Murray-Bligh and Griffiths, 2022).

Just to remind ourselves, BMWP kick-sampling scoring is primarily for use in relatively small streams and rivers. Where the river is relatively deep but narrow, a long-handled pond net can be used for sampling. It is not possible to apply this kind of sampling system for monitoring biological water quality

---

4 An additional measure also used with BMWP is the total number of scoring taxa in the sample. This metric is not shown in Table 4.

5 The symbol < indicates 'less than'; the symbol > indicates 'greater than'.

in large and deeper lakes and large rivers. In that situation a more sophisticated piece of kit – an airlift sampler – is used and normally operated from a boat. This is essentially a suction tube that sucks invertebrates from the river or lake bottom. Standard practice is to collect three-minute samples, as with the kick sampling.

The communities of invertebrates will be different in still water compared with flowing water. Still waters, which generally have lower water-oxygen levels, will be missing taxa found in well-oxygenated flowing water. However, in all these examples a decline in water quality is indicated by a change – usually a fall – in BMWP scores over time, rather than by comparing metrics from different water bodies. This decline may be due to organic sewage pollution, although there may also be other causes.

Fish biotic indicators of water quality

Where rivers or lakes are large with deep water, surveys of fish species can help to indicate water quality. We have already alluded to this in relation to the River Thames (Chapter 5), where the gradual increase in both fish species and abundance correlated with improvements of water quality. One feature that would appear to favour fish when exposed to adverse water quality is their ability to swim away from the area. In fact, invertebrates also have some ability to move through drifting with the current – referred to as invertebrate drift. Nevertheless, fish can be choosier about the direction they swim and so can more readily exit areas of poor water quality.

The types of fish that are potentially present in a clean river can depend on a variety of ecological factors. As in the case of invertebrates, some fish species are particularly sensitive to oxygen levels in the water and are found only in flowing water. Two examples of this are diadromous fish species that migrate from marine to freshwater in the UK – namely salmon (*Salmo salar*) and trout (*Salmo trutta*), which have high oxygen requirements. They will be impeded in their migration by any stretches of river where organic pollution (or other causes) lower dissolved oxygen levels. On the other hand, crucian carp (*Carassius carassius*) – occurring in the UK and northern Europe – are bottom-feeding fish in muddy and silty lakes. They are very tolerant of low oxygen levels and can survive in conditions in which oxygen is essentially absent (anoxic). Remarkably, they can use anaerobic respiration when oxygen is limiting – the explanation of which will be covered in Chapter 9.

Chapter 8
How Does Sewage Pollution Affect the Biological Health of a River?

Now that we have seen something of how water quality can be assessed biologically using indicator species, we move on to a brief look at how this links in with sewage pollution. What does the release of organic sewage do to a river? The answer to this question also connects with subsequent chapters, where we look specifically at sewage treatment and the release of treated effluent water, and sometimes raw sewage, from sewage treatment plants (Chapter 11).

Figure 7 is a rather stylized representation of what happens when a source of raw organic sewage flows into a river. Of course, many sewage outfalls may also include serious inorganic or organic chemical pollutants that can be toxic, but here we are only dealing with the organic sewage component – primarily what goes down the toilet or emerges from domestic kitchens.

We will start by considering Graphs A and B together. Note that the levels and trajectory of all the lines plotted on the graphs change dramatically at the point of sewage discharge.

Sewage, typically mixed with rainwater, arrives at an outfall loaded with many bacteria that are already busy decomposing the organic material. This process of decomposition can involve oxygen-requiring (aerobic) bacteria and anaerobic bacteria that carry out decomposition (more slowly) in environments where oxygen is absent (anoxic). The balance between these two groups of bacteria will shift depending on oxygen levels in the sewage stream.

This means that as soon as the sewage hits the river there will be a sharp drop in dissolved oxygen in the water. The aerobic bacteria, hungry for oxygen, strip the water of that precious gas. This is further confirmed by the line on the graph labelled as BOD – biochemical oxygen demand, which is the demand for oxygen by aerobic bacteria decomposing (oxidizing) the organic matter in sewage water. Oxygen is essential to drive their metabolic machinery at the cellular level, in the process of cellular respiration. There is more reference to this in Chapter 9.

To emphasize what a burden untreated organic sewage puts on a river, the organic waste produced by one person per day would require the amount of oxygen dissolved in 10,000 litres/2,200 UK gallons of fully oxygen-saturated water to complete its aerobic decomposition.

Graph A also shows the inevitable presence of suspended organic solids inflowing with the sewage water. These solids are important not only because they are organic material that will require oxygen for decomposition in the river, but also because they can have a smothering effect on animals and plants. This may impede oxygen exchange in invertebrates with delicate gills. All three measures change in concentration as the sewage water flows downstream and is diluted and re-aerated in a large volume of river water. The suspended solids may settle out on the riverbed, where decomposition continues – albeit now in an aerobic environment.

What about Graph B? This shows the elevated levels of nutrients released from the bacterial decomposition of the organic sewage. Such nutrient enrichment of natural waters is referred to as eutrophication. Nitrogenous compounds in the sewage (for example, urea – $CO(NH_2)_2$) are initially decomposed by bacteria into dissolved ammonia (labelled as NH_4). As oxygen levels rise in the mix of river water and diluted sewage, the ammonia is oxidized to nitrate (NO_3). Phosphate (PO_4) is also released from the decomposing organic sewage. The concentrations of all three nutrients fall further down the river – partly because of further dilution, but also through possible uptake by aquatic plants and algae. Plants in the river can respond with increased growth to this 'feast' of nutrients, which would otherwise be scarce in clean river water.

Moving on to Graphs C and D, these show the ecological effects that might be expected from a number of organisms that form the biota of the river. These organisms are some of the key indicator taxa that may be present in rivers at the sewage outfall but then change further downstream as sewage pollution declines from the point of discharge through dilution, decomposition and nutrient uptake. Pollution-tolerant species will be present in sewage-affected stretches of river, while many pollution-intolerant species expected to live in clean water will only appear further down the river. You can remind yourself of this by referring back to the discussion of BMWP scores in Chapter 7.

Another line on Graph C shows the distribution of sewage fungus. This is something of a misnomer because the 'fungus' is not fungi but filamentous bacteria forming a fuzzy growth on surfaces in the river (such as stones) where there is organic sewage pollution. Although not necessarily harmful, its presence does suggest release of sewage in a river. Graph D shows that small relatives of earthworms, often referred to as Tubifex worms (family Tubificidae – also known as sludge worms), can be present in huge numbers where there is organic sewage to be consumed at the end of a sewage discharge pipe. These reddish worms possess the respiratory pigment haemoglobin, also found in human blood, which mops up the low levels of oxygen and helps the worms survive the very low oxygen (close to anoxic) conditions in concentrated sewage at the outfall. As the water flows further downstream, midge larvae (*Chironomus* spp. – insects) and water hoglouse (*Asellus aquaticus* – crustaceans) are also tolerant of relatively low water oxygen levels, but largely disappear as the organic sewage dissipates and the clean water invertebrates take over. These types of changes in the river invertebrate biota are reflected in the changing BMWP biotic index score for samples taken at various downstream sampling points. This helps to establish the extent of sewage impacts at multiple distances along the river.

Algae such as *Cladophora* grow in response to the elevated nutrient levels in the water before declining further downstream. Not shown in the graph is the response of 'water weeds' (aquatic macrophytes), which are aquatic flowering plants. They can either be emergent (growing along the edges of lakes and rivers) or submergent (growing under water). An example of the latter is species of water crowfoot, classed in the genus *Ranunculus* (or *Batrachium*), which can grow extensively in flowing water courses once nutrient levels have reduced somewhat – that is, downstream of the *Cladophora*/algae zone on the graph (Figure 7). Their growth is enhanced by moderate levels of nutrients (so they are referred to as a mesotrophic species). Beyond the *Cladophora* zone, conditions also become more benign for fish that generally will not tolerate low levels of oxygen or cope with the toxicity of ammonia. Often the first fish to appear closest to an outfall are sticklebacks (*Gasterosteus aculeatus*), which are relatively pollution-tolerant.

What has been described here is a broad-brush picture of the impact of raw organic sewage and its decomposition products on a river. But this is useful background information to bear in mind when we move on to discuss sewage/wastewater treatment and the outflows from sewage treatment (Chapter 11). In a modern treatment plant such outflows should be relatively clean, but we know that sewage does still flow into rivers through CSOs at times of heavy rainfall or sewage plant malfunction.

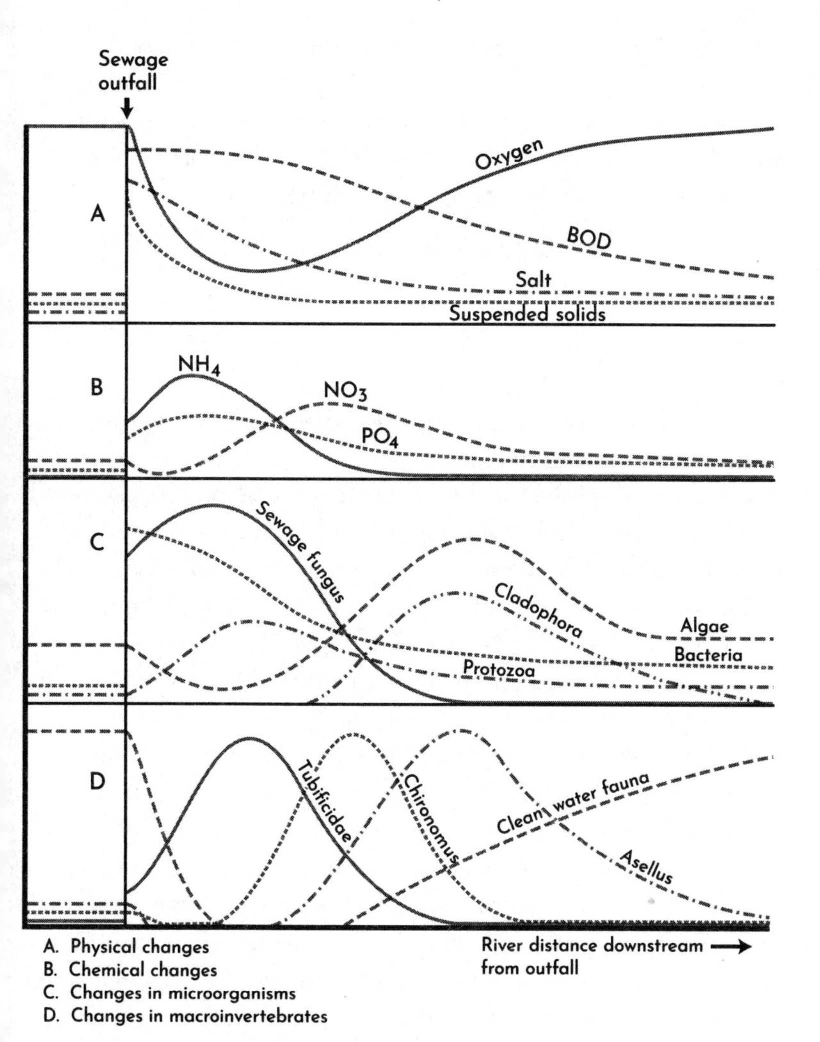

Figure 7. Graphs showing the kinds of changes in water quality and populations of organisms that might be expected in a river below a discharge of organic (sewage) effluent (Hynes, 1960).

Key: Graph A: BOD = biochemical oxygen demand; Graph B: NH_4 = ammonia, NO_3 = nitrate, PO_4 = phosphate.

Chapter 9
Sewage Treatment: The Wonders of 'Activated Sludge'

Now that we have some historical and biological background to sewage and sewage pollution, we move on to the mechanics of how noxious sewage can be treated in a way that allows cleaned sewage effluent to be returned to a river, lake or the sea. The discussion gets more technical from now on – microbes, tanks and pumps – but it should begin to make sense if you stay with it. You will then be an informed observer of every sewage treatment works you pass on the road or by train – confident that you can make some erudite comment about its functioning!

The term 'activated sludge' doesn't sound very thrilling, yet it represents the basis of much of current wastewater treatment around the world. There are newer ways of providing the infrastructure for microbial populations to break down organic matter in sewage, but they all link to activated sludge.

To begin with, let's clarify that the terms 'sewage' and 'wastewater' are used interchangeably in this chapter and book. If we were being pedantic, we would say that sewage is primarily excrement generated by people and originating from toilets; while wastewater is a more general term for all types of 'dirty' water that needs to be treated before being returned to the general environment. In reality, both can refer to inflows of polluted water into a sewer system from domestic, commercial and industrial sources. In combined sewer systems, there will also be inflow of relatively clean surface drainage water.

So, the main emphasis in this chapter is on the treatment of organic waste, primarily originating from domestic environments. Much of this is human waste from toilets – urine and faeces – but there is also some organic waste from kitchens sinks, dishwashers and so on and wastewater from baths and showers. The latter is so-called 'grey water' – which doesn't originate from toilets, but is typically loaded with soaps and detergents as well as some organic material. To keep things simple (or simpler!), we will avoid the complications of wastewater from commercial/industrial premises. Typically, industrial processes that generate significant amounts of concentrated organic waste (for example, dairy processing) are required to have onsite pretreatment of wastewater to reduce the load on the sewerage system.

Before we start delving into the main stages of 'cleaning' sewage, we will cover some biological information that is needed to explain what occurs in a wastewater treatment plant. This mainly concerns what bacteria get up to when confronted by sewage. These key aspects of the basic biology of bacteria are explained next.

Bacterial respiration

Bacteria are crucial 'little helpers' in sewage treatment. The effect they have on the breakdown (decomposition) of sewage organic matter is a consequence of bacterial respiration. In relation to humans, respiration tends to be thought of as 'breathing', but in this context its meaning is more focused. Respiration is a process that occurs in all cells, including microbial cells. It refers to the biochemical (metabolic) breakdown of energy-rich molecules to release energy used to drive the biochemical machinery of the cell. In human terms, this would equate to taking in a food substrate such as the carbohydrate sugar and then breaking it down in cells

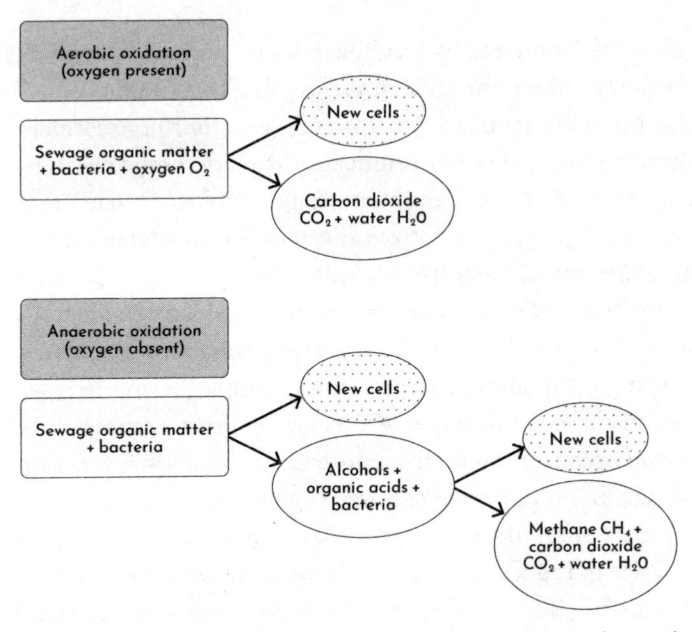

Figure 8. A simplified representation of aerobic and anaerobic oxidation of organic matter in sewage. Ammonia and hydrogen sulphide are additional products of anaerobic respiration.

to release energy that runs muscle contraction when moving and a whole host of cell and tissue functions. Human cells require oxygen to complete the breakdown of sugar – with the subsequent release of carbon dioxide as a waste product. This requirement is met by breathing: oxygen in and carbon dioxide out. By contrast, as we will see, not all bacteria require oxygen for cellular respiration.

Bacteria break down (metabolize) organic waste in sewage to provide energy. But unlike humans, bacteria use one of two forms of cellular respiration: aerobic (which requires oxygen) or anaerobic (which doesn't) – as shown in Figure 8. You will note that anaerobic respiration is incomplete and so yields less energy per unit weight of organic matter. Instead of carbon dioxide – which has no usable energy – organic alcohols, acids and methane are products of anaerobic respiration and still have an energy content. This is evidenced by the fact that methane and alcohol will burn and release energy primarily as heat. Ammonia and hydrogen sulphide (both pungent and potentially toxic gases) are additional decomposition products of organic molecules under anaerobic conditions.

Sewage and oxygen

Huge volumes of wastewater are constantly passing through the sewerage system to the closest sewage (wastewater) treatment plant. Some anaerobic decomposition will begin within the sewerage system even where oxygen levels are low. However, the bulk of sewage treatment requires the uptake of plentiful oxygen. We can gauge the 'strength' of that sewage by measuring its biochemical (or biological) oxygen demand (BOD) – how much oxygen is required by the bacteria in the sewage to decompose the organic content of the sewage. A higher concentration of organic waste in the sewage equates with a higher BOD value.[6] Domestic sewage has a typical BOD value of 250–400 mg/l (or mg l^{-1}). That doesn't seem a lot – less than half a gram of oxygen for every litre of sewage. But (and this is a big 'but') the amount of oxygen that can dissolve in water (by weight) is far lower than it is in air. Oxygen represents about 23 per cent of the weight of a specified volume of air. This means that a litre of air will contain about 297 mg of oxygen (it varies with temperature and air pressure), whereas a litre of water at 21°C/69.8°F contains only around 8.7 mg of oxygen (this also varies with temperature and pressure) – so only about 3 per cent of the value (weight) in air. This means that the typical BOD value of domestic sewage, as quoted above, would far exceed the amount of oxygen that could be dissolved in the sewage water.

6 It is also possible to express the organic content of sewage using a measure called chemical oxygen demand (COD). This gives a higher value than BOD because it includes organic material that is not readily biodegradable by bacteria. As a laboratory test it has the advantage of being faster to perform than BOD.

Any oxygen that is present in sewage will be quickly stripped by bacteria. This means that sewage can rapidly become totally anoxic – although some (very minimal) replenishment occurs through turbulence as it flows through the sewage pipe system. We have already noted that only anaerobic bacteria will operate in the absence of oxygen, so sewage decomposition in these conditions will be not only slow but also incomplete. This is something to bear in mind when we consider what needs to happen when the sewage reaches the treatment plant.

The ultimate aim of sewage treatment in the UK is to reduce the BOD by about 90–95 per cent, to a maximum of 20 mg/l (mg l^{-1}), before treated water is returned to the river.

In the following section we will go through the main stages of the sewage clean-up process. This involves a mix of physical, biological and chemical methods for converting highly polluting sewage influent into final, relatively clean effluent.

Stages of wastewater treatment

We start the story of wastewater treatment by presupposing that we are dealing with domestic sewage that has flowed through the sewerage system to the treatment works.

Sewage consists primarily of water – up to 99.9 per cent. Only about 0.1 per cent is solids (both organic and inorganic) in suspension. Nevertheless, sewage represents a rich store of nutrients – millions of metric tonnes on a world scale. Urine accounts for 50 per cent of phosphorus and 80 per cent of nitrogen in sewage (Bhat and Qayoom, 2022); the rest is in the faeces.

The processing used at sewage treatment plants can seem complex, but it is important to stress that it involves only three basic stages. These stages do the bulk of the work in removing organic matter from sewage and reducing BOD.

They are:

Stage 1: Primary settlement/sedimentation
Stage 2: Activated sludge treatment in reactor tanks
Stage 3: Secondary settlement/sedimentation

Of course, it's not *quite* that simple. This basic set-up involves some additional stages that are ticked off in sequence as follows:

Preliminary stage

✓ Screening treatment (physical): removal of larger solid waste – *simple*

Stage 1

✓ Primary settlement (physical): first settlement/ sedimentation – *simple*

Stage 2

✓ Activated sludge treatment (biological): activated sludge in reactor tanks – *complex*[7]

Stage 3

✓ Secondary settlement (physical): final settlement/ sedimentation – *simple*

Concluding stage

✓ Disinfection (physical or chemical): chemical agents/ UV/filtration – *simple*

There are also . . .

✓ **Alternative biological** treatment options (Stage 2): described in Chapter 10 – *complex*

7 An additional stage of **nutrient removal** as part of **activated sludge treatment** (Stage 2) is discussed later in this chapter.

A schematic diagram of how these various stages fit together as a continuous flow system is shown in Figure 9.

Preliminary stage: screening treatment

In an ideal world, sewage arriving at a sewage treatment plant is a 'soup' of small organic particles in suspension that pass easily through the rest of the treatment system. However, a variety of more solid objects also find their way into sewage. Apart from toilet paper, which disintegrates quite easily, there will be other materials, such as condoms, menstrual products and, these days, the ubiquitous wipes – which typically have a significant plastic content. Plus anything else people choose to flush down the toilet or that gets dropped in by accident! Due to water inflows from road surfaces, combined sewage systems also typically contain a significant amount of grit. This means that the sewage needs to pass through metal screens to sift out miscellaneous solid material that could otherwise clog the many pumps required to move sewage around the plant. Mechanical scrapers are used to remove this noxious mix of solid waste, which will normally go to landfill for disposal. Grit is removed in the grit settlement tank.

The next three key stages in removing organic waste are shown in Figure 10.

Stage 1: primary settlement/sedimentation

After screening, the sewage flows on to the first of the three key stages that do the bulk of the work in reducing the organic content of sewage. This first 'proper' stage of treatment aims to substantially reduce the BOD of influent sewage through the physical process of sedimentation – known as primary settlement. Slowing down the speed of sewage flow causes the heavier solid waste to settle out while the rest continues

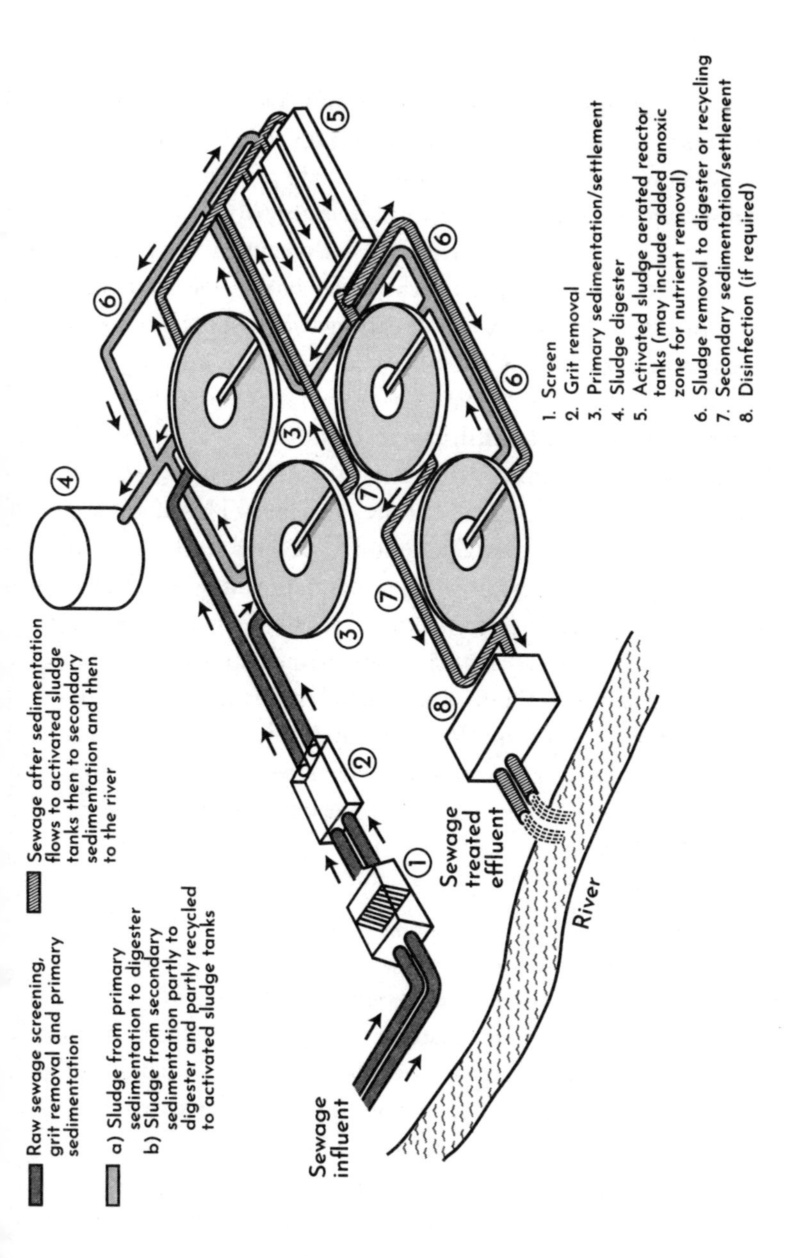

Figure 9. A schematic diagram of the main parts of a sewage treatment plant based on activated sludge treatment. Influent sewage passes through stages of treatment before clean effluent is discharged to the river.

to the next stage. Around 90 per cent of the heavier settleable solids and 50 per cent of the lighter suspended solids are taken out of the sewage flow in this way. The net effect of this first stage of processing is to reduce the BOD of the sewage by about 25–40 per cent. The settlement tank is equipped with scrapers that periodically discharge the raw sewage sludge for further treatment (sludge digestion). This will be discussed in Chapter 10.

Stage 2: activated sludge treatment

It is at the next stage of sewage treatment in the aerated reactor tank where the biological 'magic' happens. This stage is prosaically called the activated sludge process and is the most crucial part of this wastewater treatment story. To give credit to the originators of this process, let's start with a bit of history. The method was first described in a paper presented at a meeting of the Society of Chemical Industry in Manchester by Edward Ardern and William Lockett (1914). They are little known outside the sewage industry, yet the process they invented has become a mainstay of sewage treatment throughout the world – albeit in several different guises. So, we have here two unsung heroes!

At Stage 2 the biological (rather than physical) treatment of sewage begins in earnest in an aerated, oxygen-rich (aerobic) environment. It's important to remember here that the decomposition of organic material (oxidation) requires an assemblage of (mostly) aerobic bacteria, which need to be provided with optimum conditions for them to operate. 'Feeding' them is the organic matter in the sewage, which is a source of both energy for bacterial cells and of building materials (nutrients) to make more bacteria. The key condition to keep these aerobic bacteria happy and working hard

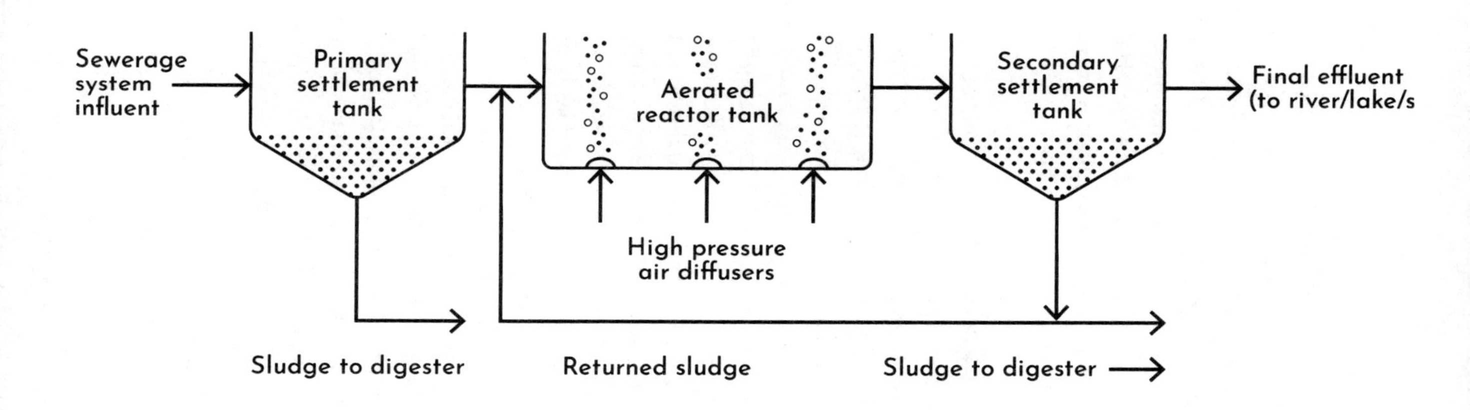

Figure 10. Sequential stages of treating wastewater involving activated sludge in a reactor tank.*

*Additional treatment is not included in this figure, which represents only the basic components of activated sludge treatment.

is, of course, the supply of oxygen – which drives the bacterial oxidation of the organic matter.

There are several designs of treatment plants that can create these conditions, but we will first focus on a basic design that has been very widely adopted in sewage treatment. It continues in use in many treatment plants but there are also newer, more efficient designs based on the same principle of aerobic bacterial decomposition. A few of these will be discussed in Chapter 10.

WHAT IS ACTIVATED SLUDGE?

Sewage that has passed through primary sedimentation/settlement flows into rectangular tanks that are typically referred to as reactors. These can be very large if they are processing enormous volumes of sewage in a major city. This is a continuous flow system – influent at one end and effluent at the other end of the tank. The tanks harbour a vast population of numerous bacterial species, as well as other small unicellular and multicellular organisms. Collectively these organisms can be thought of as a microbial ecosystem, with a variety of functions and roles ascribed to the different microbial and other species. As is true of ecosystems in general, a food web is in operation, with waste organic matter driving the system of bacterial species (decomposers) 'feeding' on the waste while other species predate the bacteria. These predatory organisms include the single-celled Protoctista (protozoa) and multicellular organisms like nematodes (roundworms) and rotifers (wheel animals) (Figure 11). In view of the current enthusiasm for the term 'gut microbiome', this could be referred to as a sewage microbiome.

The living organisms that inhabit the activated sludge may float about freely in the sewage liquor, but they are concentrated within floc. Floc is a loosely clumped mix of delicate cotton-wool-like organic colloidal material (complex polymer carbohydrate molecules) produced by the microorganisms. With the addition of other inorganic components, the floc stays in suspension and provides a home (matrix) for a host of organisms living within the floc. Hence, it is the floc organisms that provide biological-processing power and drive the breakdown of organic sewage material. As we will see, in some other types of wastewater treatment, this community of organisms operates as a biofilm attached to a surface rather than as free-floating floc.

We can now say more about what exactly occurs in a tank full of sewage and microorganisms. When this type of reactor tank is in action at a sewage plant, there will be very noticeable and constant streams of bubbles breaking the surface of the sewage mix (liquor). These come from air continuously pumped under pressure through diffusers at the base of the tank – not only to mix the tank's contents but also, most importantly, to force oxygen to dissolve in the sewage liquor. This supercharges the microbial population so that the process of oxidation of the organic matter can begin in earnest. The initial high BOD of the mix is accommodated by the vigorous aeration, which causes rapid oxygen replacement as it is utilized by bacteria. The rate of oxygenation must match that of deoxygenation to prevent anaerobic decomposition. It is this injection of air into the flowing sludge that provides the moniker 'activated sludge'.

THE ROLE OF LIVING ORGANISMS

The bulk of the decomposition process in the reactor is linked to the work of bacteria – of which there are likely to be more than a dozen genera and many species and strains. Different components in the sewage are decomposed by different bacteria – that is to say, there is a degree of specialization. Some examples of this are shown in Table 5.

Bacterial genus	Role in sewage treatment
Achromobacter	Denitrification
Acinetobacter	Phosphorus removal
Bacillus	Protein breakdown
Nitrobacter	Nitrification
Nitrosomonas	Nitrification
Pseudomonas	Breakdown of carbohydrates and denitrification

Table 5. Examples of genera of bacteria found in activated sludge and their role in relation to the decomposition of organic sewage.

What about the single-celled Protoctista (protozoa) shown in Figure 11? The label 'protozoa' is still commonly used for single-celled higher organisms (that is, not bacteria), but it has officially been dropped as advances in taxonomy now place these organisms in the kingdom Protoctista. Even that label has been superseded, but we will stick with it for convenience.

The rhizopods (amoebae), ciliates and flagellates are three groups of protoctists that are found in sewage. The ciliates are the most important, both numerically and for their contribution to sewage treatment. They are small – up to about one-third of a millimetre in length, but many are much smaller. Ciliates have many cilia – tiny, short, motile 'hairs' that provide propulsion for them to glide through the watery medium they occupy. In favourable conditions there may be 20,000 to 100,000 ciliates per cm^3 (or 1 ml) of activated sludge and they may constitute 4–5 per cent of the dry weight of suspended solids. They contribute to capturing and consuming bacteria, including disease-causing (pathogenic) species.

Small aquatic multicellular organisms are also present in activated sludge (Figure 11). Their ecological role is also that of consumers of bacteria. Because they are so much larger than

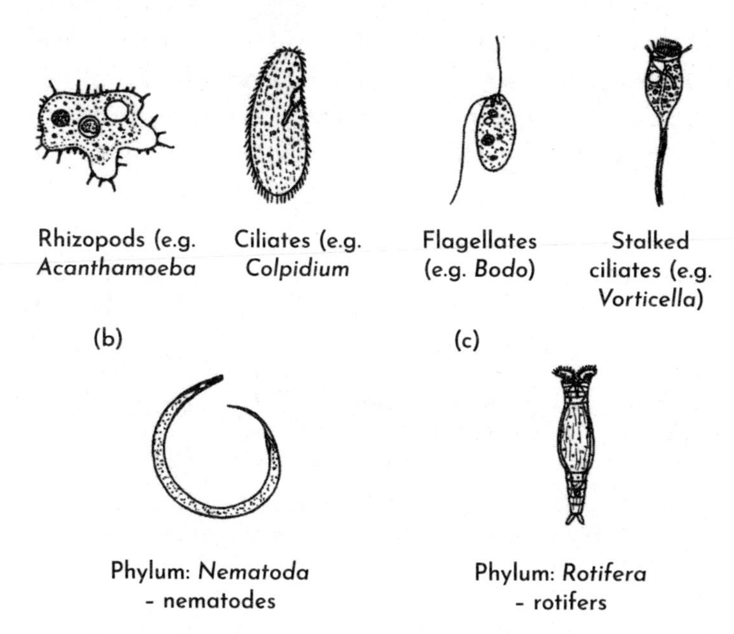

Figure 11. Typical invertebrates of activated sludge: (a) unicellular and (b) multicellular. Larger species may be about 1 mm in length, but others are considerably smaller. Most would be classed as microscopic.

protoctists and bacteria, they are also more apparent if you examine sewage floc under a microscope. One group, the rotifers, are typically about 0.2–1 mm in length. They have the common name of wheel animals because they have paired wheel organs at one end – used for filtering and consuming suspended particles and bacteria from the sewage mix. Nematodes (roundworms) can also be very numerous and similarly feed on suspended particles and bacteria.

A mix of unicellular and multicellular organisms will be present throughout the reactor tank, which contains a mix of fresh sludge/floc as well as older (aged) floc that has spent time in the reactor tank due to recycling of sludge, as shown in Figure 10. The aged sludge/floc will have a different biota than the fresh sludge/floc. There is a process of ecological succession of organisms as sludge/floc ages, 'early' species being replaced by 'later' ones as conditions in the sludge/floc change (Figure 12).

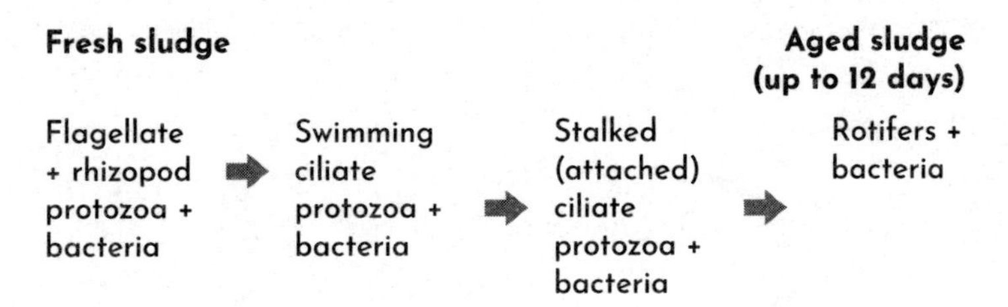

Figure 12. Successional change of Protoctista and multicellular organisms with age of sewage sludge/floc as it passes through treatment and conditions change to suit different species.

OXIDATION OF SEWAGE

Now that we know the 'actors' in the sewage treatment 'drama', we can try to address the question of what changes occur as the activated sludge passes through the reactor tank.

The main change is the oxidation of organic matter in the sewage. Like all organic matter, this material is rich in complex organic carbon-based molecules that are characteristic of living organisms and of the food consumed by people and other animals. For example, plant material will be rich in cellulose, a long-chain polymer that cannot be digested in the human gut and is voided as faeces. On the other hand, many bacteria can break down cellulose because they possess cellulase enzymes to snip the cellulose molecules into smaller 'sugar' molecules that can be used as a source of energy. Aerobic bacteria need oxygen to oxidize the carbon content of the organic material and in doing so exploit these molecules as a source of energy to drive their cellular processes. This is also what happens when humans break down sugar aerobically: there is a release of energy and the oxidized carbon is released as CO_2 (carbon dioxide). So, as sewage flows through the reactor tank, there is a steady reduction in the carbon content of the sewage – it disappears into thin air as carbon dioxide gas. This presents a delicate balancing act for treatment plant operators. If the carbon content falls too low for the population of bacteria, then the bacteria starve! The reactor tank needs to maintain a high population of active bacteria. This balance between bacteria and 'food' is expressed as the F/M ratio (food to microorganisms). When the F/M ratio is high, bacteria are active and multiply rapidly. The amount of bacterial 'food' can itself be expressed in terms of a BOD value (see Chapter 8). Sewage with a high concentration of organic material, or 'food' for bacteria, will have a high BOD.

Of course, carbon is not the only element present in the rich mix of molecules in the sewage. Organic molecules like proteins and nucleic acids (DNA) contain elements such as nitrogen and phosphorus. These elements are particularly important to our story of sewage processing and we will return to them shortly. For the present, suffice to say that both these elements will be released in soluble form in the reactor tank during bacterial decomposition.

Stage 3: secondary settlement/sedimentation

Earlier in this chapter we referred to primary settlement/sedimentation at the start of wastewater treatment to reduce the quantity of settleable/suspended (mainly organic) solids. After sewage has passed through the activated sludge reactor tanks, there will still be some solids in suspension – including bacterial floc. These are substantially removed by the final stage in Figure 9, secondary sedimentation/settlement. According to UK requirements, at Stage 3 the quantity of suspended solids must be reduced to no more than 30 mg/l (30 mg l^{-1}) for the release of 'clean' effluent water.

As in the primary-settlement tank, some of the settled sludge is removed for further treatment. But there is one important twist to this part of the story: part of the sludge from secondary settlement is recycled back to the activated sludge reactor tank. Why? Well, if all that sludge was lost, it would take with it much of the microbial and other organisms that are the 'engine room' of sewage processing; a lot of the floc will be in the sludge. Partial recycling therefore maintains the reactor tank microbiome.

So far, so good – until the problems start. Chief among those problems is the dreaded 'bulking sludge'. This is a reminder that this is a biological system that can sometimes veer off-track – because it's not always possible to control fully an intricate microbial ecosystem with variable sewage input. What is bulking sludge? Sludge bulking occurs when sludge doesn't sink to the bottom in the secondary settlement tank. This is the point when the floc, no longer being agitated, should sink to the bottom. The reason for the bulking is the extensive growth of filamentous bacteria that have different surface area characteristics and settle out only slowly. There may be other conditions, too, such as rising sludge caused by the production of bubbles of nitrogen gas, which lifts the sludge to the surface. The presence of nitrogen gas is the result of denitrification, covered in the next section. The life of a sewage treatment plant manager is not an easy one! The effect of both of these problems is a reduction in the quality of sewage treatment, poorer final effluent quality and possible odour issues.

Nutrient and ammonia removal

Activated sludge treatment does normally eliminate the most noxious elements of sewage. The organic sewage material in suspension has largely gone and after secondary sedimentation the effluent looks fairly clean. There will also be a major reduction in pathogenic bacteria. However, the dissolved components in the final effluent mean that this is not the end of the activated sludge story. We mentioned earlier that soluble forms of nitrogen and phosphorus are released during the decomposition of sewage. These solutes remain in the effluent and so will flow into a river, lake or sea. One of these solutes is ammonia, which is toxic and for which there are permit limits for release in final effluent. Generally, ammonia must be substantially removed during wastewater treatment. Other solutes are nutrients that will cause nutrient enrichment of water (eutrophication) and hence a problem of excessive plant and algal growth – as discussed in Chapter 8. Finding a way to eliminate these 'invisible' nutrients and toxic ammonia as part of activated sludge treatment takes us into the complex realms of microbial metabolism and the chemical transformation of molecules and ions. So, the following explanation is somewhat convoluted but represents a rather clever trick of using bacteria to remove nutrients and ammonia with a bit of extra engineering, pumps and pipework.

The nutrient-removal stage is alternatively known as the 'tertiary treatment' stage. Whichever term is used, the process relates to nutrient removal from final effluent before it is released into a water body. There is a requirement for nutrient removal/ tertiary treatment in the UK where the following conditions apply:

> You must use more advanced tertiary treatment for agglomerations of more than 10,000 PE (Population Equivalent) in sensitive areas and their catchments

> where the discharge contributes significantly to nutrient load. (Environment Agency, 2023)

We listed the stages of wastewater treatment at the start of this chapter. We can now add the extra nutrient-removal stage (in bold) as an extension of the activated sludge process.

Preliminary stage

✓ Screening treatment (physical): removal of larger solid waste – *simple*

Stage 1

✓ Primary settlement (physical): first sedimentation – *simple*

Stage 2

✓ Activated sludge treatment (biological): activated sludge in reactor tanks – *complex*

✓ **Nutrient and ammonia removal (biological or chemical): modified activated sludge treatment – *complex***

Stage 3

✓ Secondary settlement (physical): final sedimentation – *simple*

Concluding stage

✓ Disinfection (physical or chemical): chemical agents/UV – *simple*

AMMONIA AND NITRATE REMOVAL

To understand how nitrogen is removed in the activated sludge process, we need to understand the somewhat complex transformations of nitrogenous molecules that occur in sewage. Nitrogen will be present in sewage in several forms, both soluble and in solid organic material in suspension. This will include large, insoluble, organic protein molecules and soluble organic urea ($CO(NH_2)_2$). Inorganic ammonia (NH_3), ammonium ions (NH_4^+) and inorganic oxides of nitrogen (NO_2^- and NO_3^-) will be present in solution. There is interchange and transformation of these various forms of nitrogen during sewage treatment, as shown in Figure 13. Note that nitrogen fixation is part of the nitrogen cycle in most ecosystems but is essentially not relevant to our discussion – hence the brackets around that label. There is also a new term here: assimilation. This is the take-up of nitrogen into organisms for incorporation into cells as they grow and reproduce. This will apply, for example, to microbial cells growing and multiplying in the activated sludge, which need to take up nitrogen to incorporate in new cell proteins.

One starting point for nitrogen conversion is the hydrolysis of urea molecules (the major nitrogen source in urine) to form ammonia. But this is not the only source of ammonia. We know that all living organisms contain proteins that will be present in faecal material as well as in the microorganisms populating the sewage. Proteins are disassembled by various bacteria to release ammonia in both oxygen-rich and oxygen-poor sewage. The microorganisms extract energy from this process – known as ammonification (Figure 13). Ammonia is relatively toxic in solution so the next step in sewage treatment involves oxidation of ammonia to much less toxic nitrite and nitrate ions. Two genera of bacteria, *Nitrosomonas* and *Nitrobacter*, are responsible for this two-stage process, which is also an energy-yielding process for these bacteria.[8] This two-stage oxidation is called nitrification and requires oxygen-rich (aerobic) conditions.

The two genera of bacteria responsible for nitrification are relatively 'fussy' organisms, which are sensitive to conditions in the sewage liquor and can only respond slowly to changed concentrations of substrate. So, they are most effective where reactor tanks build up high populations of both bacteria. Conditions in the reactor must be managed to suit the biology of the bacteria.

Although the conversion of toxic ammonia to more benign nitrate is an essential step in sewage treatment, we have previously alluded to the downside of releasing the nitrate to the environment in final effluent from sewage treatment. To reduce the release of nitrates requires a further process, known as denitrification, to convert nitrates to nitrogen gas, which is safely released into the vast pool of nitrogen in the atmosphere. This is achieved in the following way.

Nitrate concentrations in final effluent can be reduced during sewage treatment by incorporating an extra tank in the processing system (Figure 14). Sludge containing nitrate is recycled from the aerobic (Tank B) to a reactor (Tank A) where 'fresh' sewage from the sewerage system will contain little or no oxygen – an anoxic environment. In the absence of aeration in the anoxic tank, a variety of microorganisms (for example, *Pseudomonas* and *Achromobacter* spp.) have a capacity for anaerobic nitrate respiration. In doing so, nitrate is reduced through intermediate stages to nitrogen gas, which can escape from the system – the process of denitrification (Figure 13).

Let's just refer to Figure 14 again and recap what is going on at each stage. Figure 14 has obvious similarities with Figure 10,

---

8 More recently, members of the bacterial genus *Nitrospira* have been shown to have a similar nitrification capability but can perform both stages of nitrification (known as comammox bacteria) rather than just one. These bacteria are also present and active in sewage treatment.

but with the addition of a separate reactor tank (A) placed before the main aeration tank (B).

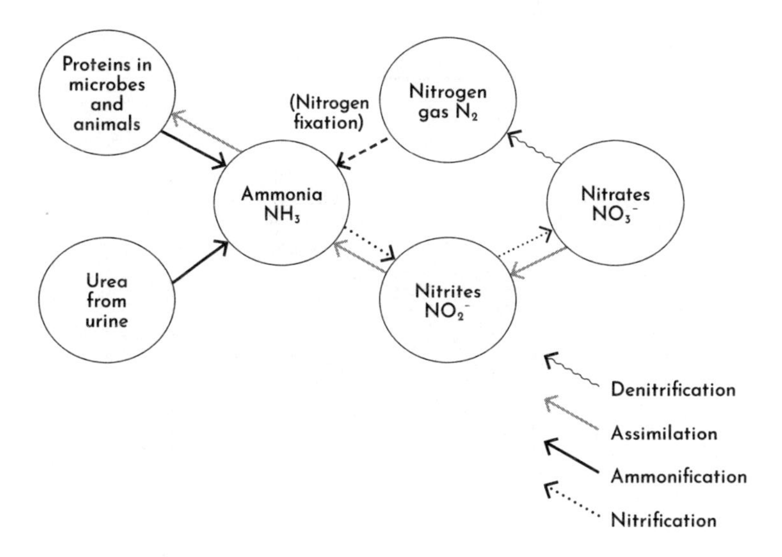

Figure 13. Simplified nitrogen cycle showing the transformations that occur during sewage treatment. Note that nitrogen fixation will not normally occur during activated sewage treatment.

Stage 1: ammonification. Sewage arriving at a treatment plant will already have some dissolved ammonia. After primary settlement, there is further conversion of organic nitrogen to ammonia (NH_3 or NH_4OH if dissolved in water) by ammonification in Tank A. This happens in the no-oxygen (anoxic) environment of Tank A, where there is no injection of air. So, sewage arriving in the aerated Tank B contains dissolved ammonia.

Stage 2: nitrification. In the presence of abundant oxygen in the strongly aerated activated sludge reactor B, ammonia is

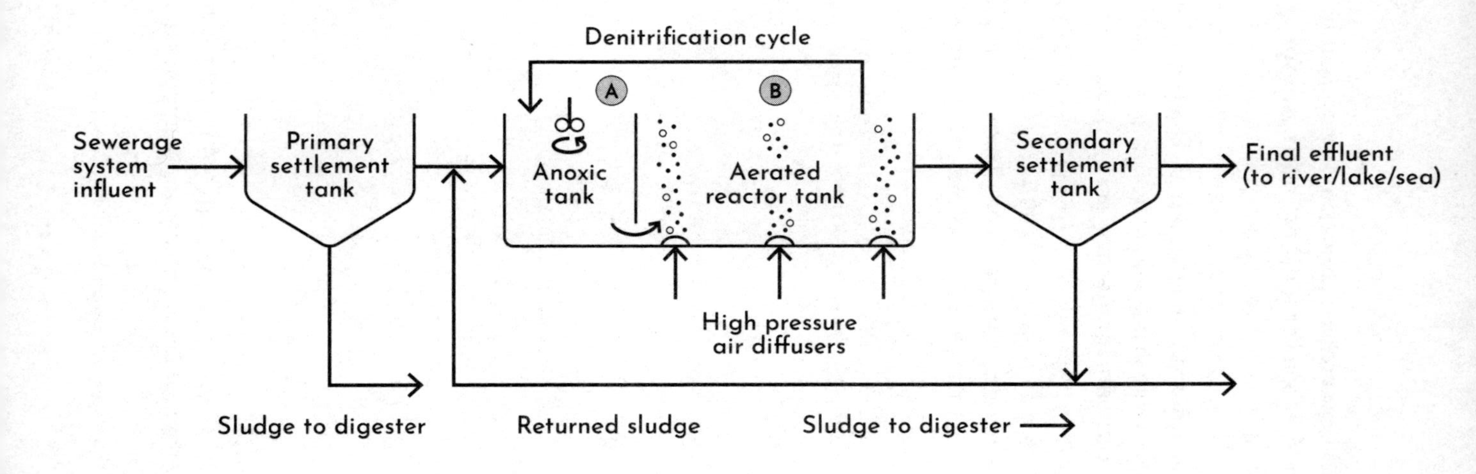

Figure 14. Sequential stages of treating wastewater involving activated sludge in a reactor tank – but modified for nitrogen removal.

first oxidized to nitrite ions (NO_2^-) by *Nitrosomonas* bacteria, then further oxidized to nitrate ions (NO_3^-) by *Nitrobacter* bacteria (Table 5). The sewage effluent leaving Tank B will have significant quantities of nitrogen in the form of nitrate, which needs to be removed.

Stage 3: denitrification. Instead of being pumped to the final stage of secondary sedimentation, much of the sewage liquor in Tank B is returned (recycled) to Tank A – the 'denitrification cycle'. In Tank A, which is an anoxic environment, anaerobic denitrifying bacteria (see Table 5, *Achromobacter* and *Pseudomonas*) get to work using organic carbon as a source of energy and in doing so convert nitrate to nitrogen gas (anaerobic respiration), which escapes into the atmosphere. By continuously recirculating sewage between Tanks A and B, the level of nitrate in the final effluent falls.

This is the end of one complicated story – nitrogen removal from sewage. Some nitrate will still be released in final effluent, but there should be little ammonia.

The other element that needs to be removed from final effluent is soluble phosphorus (orthophosphate ions PO_4^{3-}). This will be explained in outline and can be added here as Stage 4.

PHOSPHATE REMOVAL

Stage 4: biological and chemical phosphorus removal. Much of the phosphorus in the environment is typically in a form that is insoluble and immobile in ecosystems. But there is soluble phosphorus in sewage. The main soluble form is orthophosphate (PO_4^{3-}), although other soluble forms, such as polyphosphates (written as $(PO_4)n$), may also be present. Phosphorus in sewage derives from urine and faeces but also from laundry and dish- washer detergents. Although there are regulations restricting the amounts of phosphorus that can be added to these products, their contribution to phosphate in sewage is nonetheless significant.

So, how to remove the phosphorus? Once again, nature can be exploited in a way that is remarkably useful and convenient for us.

Table 5 refers to phosphorus removal and a genus of aerobic bacteria called *Acinetobacter.* These bacteria store phosphorus inside the cell as particles of polyphosphates (called volutin granules) that can be used as a source of energy by *Acinetobacter* under specific conditions. Several other bacterial genera have also been shown to accumulate phosphorus in this way. If a treatment plant is modified so that sewage first passes through an anaerobic/anoxic tank, *Acinetobacter* cells are forced to metabolize their stores of polyphosphates as a source of energy. In doing so they release soluble orthophosphate ions into the surrounding sewage. At that point, paradoxically, the amount of soluble phosphate in the sewage liquor actually rises. But then, as the bacteria pass into the highly aerated activated sludge zone, they can restore their oxygen-dependent metabolism and are 'hungry' to replenish their polyphosphate store.

They therefore begin to take up phosphate from their immediate environment – but, crucially, they take up more phosphate than they have previously lost. The phosphorus content of these bacteria is then about 5–7 per cent of their biomass. So, they take up not only the soluble phosphorus they have released, but also additional soluble phosphorus from the sewage. This pattern of phosphate release and uptake is shown in Figure 15, where it is referred to as luxury uptake of soluble phosphorus. This process is known as enhanced biological phosphorus removal (EBPR) and is a very convenient biological way of mopping up phosphorus in sewage. The phosphorus trapped in these bacteria will be 'siphoned off' as part of sludge removal from the system in the secondary settlement tank. Although this is a very benign method of reducing phosphorus concentrations in sewage effluent, it does need clear understanding and close control of the sewage microbiomes.

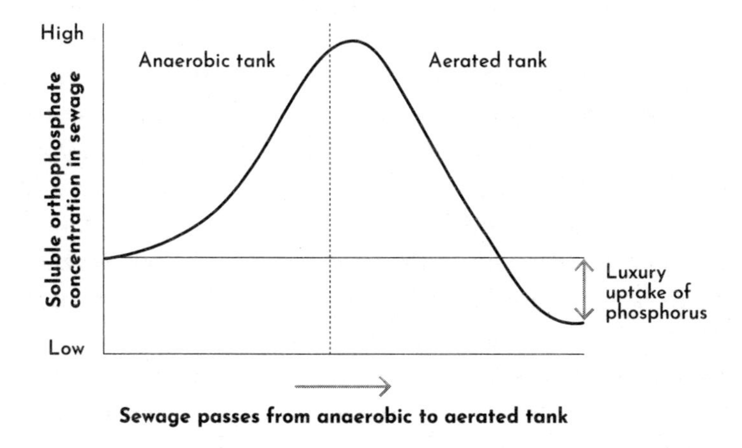

Figure 15. Luxury uptake by *Acinetobacter* bacteria of soluble phosphorus when sewage liquor is transferred from anaerobic to aerobic conditions. The concentration of orthophosphate in the sewage effluent is reduced.

Such biological phosphorus removal in an activated sludge system can work well and is eco-friendly, but it reduces rather than totally removes soluble phosphorus. Not all older treatment plants can be adapted to achieve such phosphorus removal and there are other limitations with its use. The alternative is to use chemical methods. The addition of calcium, iron or aluminium salts to sewage liquor will precipitate phosphate, which is then removed with the sludge in the secondary settlement stage. Although this method is easily controlled, it has the disadvantage of cost for the chemical additives, and precipitated phosphorus significantly increases the volume of sludge. Where there are specific limits on concentrations of soluble phosphorus in effluent, a combination of biological and chemical phosphorus removal is often used to achieve compliance.

Concluding stage: disinfection

This should be the end of the story. The effluent is as clean as it can be if passed through a modern, well-functioning sewage treatment works. Can it now safely flow into the river, lake or sea? Well, maybe. Let's say it is going to be released into a stretch of river or beach that officially comes under a UK 'bathing waters' designation. Such areas, which are mainly coastal in the UK but also include some inland waters, must meet specific water-quality standards with respect to *E. coli* and intestinal enterococci bacteria. Where there is concern about meeting these standards, sewage effluent can be disinfected as it leaves the treatment plant. There is currently little use of disinfection in the UK apart from isolated bathing waters, where ultraviolet light is typically used to disinfect sewage effluent. However, with the likely designation of more inland bathing waters in the UK, the requirement for disinfection is likely to increase.

This is a point in the wastewater treatment narrative where you could stop reading. We have already covered the basic principles of sewage treatment and the next chapter covers technical changes in plant design that lead to better-quality final effluent. Putting the book down at this stage might be a bit hasty. The advice would be to read on to learn about some of the more recent innovations . . .

Chapter 10

Variations on a Theme: Alternative Sewage Treatment Systems and Sludge Disposal

Since the activated sludge process was described about a century ago, well-oxygenated populations of bacteria have become the foundation of wastewater treatment. But the ways in which such bacterial populations are used have spawned a variety of newer technologies, some that are now forty or more years old and others that were developed more recently. These processes do a similar job in different ways and can be more efficient. That is to say, they may have a smaller footprint (occupy less valuable land) or use less energy or produce better-quality effluent. This could be summarized as producing final effluent of an appropriate standard (for the location) at a lower cost. This can easily be done when designing a new sewage treatment plant but in many cases these new technologies are retrofitted to existing treatment plants. Hence the set-up and design of these systems will vary and may be different from the descriptions in this section.

Moving bed biofilm reactors (MBBRs)

This system is an upgrade of the standard activated sludge plant already described in the previous chapter, but uses a fixed film of bacteria rather than floating floc. It involves little plastic carriers, shaped rather like cartwheels, with a diameter of about 1–3 cm (0.4–1.2 inches) across (Figure 16). Thousands of these are added to an activated sludge reactor (Figure 17). But why? They provide a surface on which a dense biofilm of bacteria develops, again with forced aeration. You can imagine that this huge volume of biofilm, along with floating bacterial floc, greatly enhances the processing power of the reactor tank. As well as augmenting the aerobic functioning of existing treatment plants, new plants can be built with a smaller footprint for the same treatment capacity. Alternative systems exist where there are fixed surfaces in the reactor tank on which bacterial films develop.

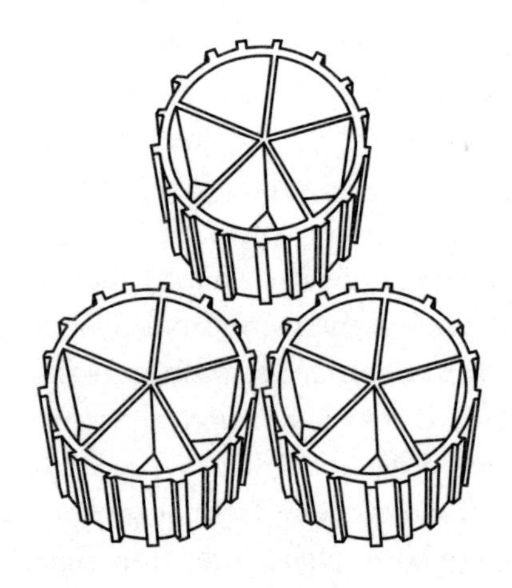

Figure 16. Plastic carriers that provide a biofilm substrate in moving bed biofilm reactors (MBBR).

Membrane-based bioreactors

The 'new kids on the block' are wastewater treatment processes that rely on membranes in one form or another, coupled with variants of the activated sludge system. Most of these methodologies are relatively recent and some have yet to prove

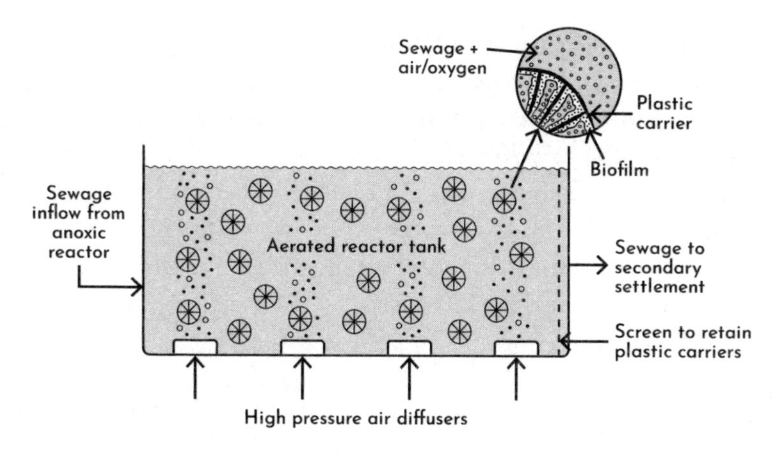

Figure 17. Moving bed biofilm reactor (MBBR), a modified aerobic activated sludge process with plastic carriers that provide a surface for growth of biofilm.*

*Plastic carriers are about 1–3 cm (0.4–1.2 inches) in diameter and are not drawn to scale.

themselves over the longer term. Nevertheless, on paper they can offer improvements in effluent quality and/or cost-effectiveness. We will look at two contrasting examples.

Membrane bioreactors (MBRs)

Imagine racks of thin-walled vertical tubes, each made of a flexible, semi-permeable membrane and held within metal frames. These frames are positioned in a reactor tank through which sewage flows after first passing through the activated sludge reactor. All the tubes in a rack join an outlet pipe (Figure 18). What do these membrane tubes do? The outside of each tube is bathed in sewage, which is mostly water. This water can selectively filter through the walls of the membrane to fill the inside of each tube. This is clean water, which flows out of the MBR outlet pipe as the final effluent. The membrane used can have a tiny pore structure (0.01 µm or less; 1,000 µm = 1 mm) so neither sewage sediment nor bacteria will pass through; the effluent is therefore very clean water. This process of ultrafiltration allows the recycling and reuse of this final effluent as clean water. Alternatively, the effective removal of bacteria makes this system particularly useful where effluent is released into bathing waters, where the MBR system can contribute to achieving the required microbiological standard.

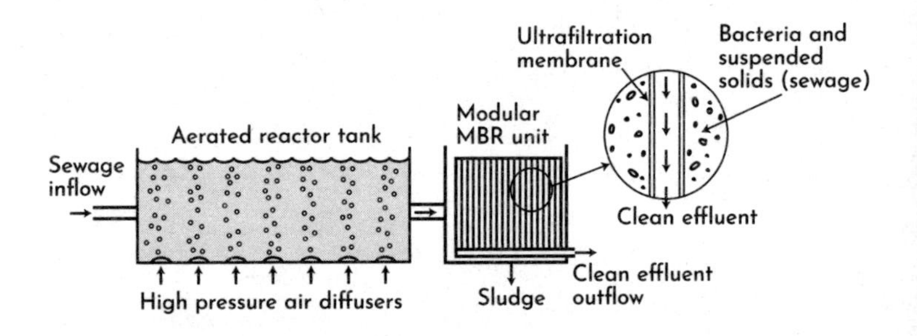

Figure 18. Membrane bioreactor (MBR). A conventional activated sludge process precedes ultrafiltration in the MBR unit, which can produce a suspended solid and bacteria-free final effluent.*

*Bacteria and suspended solids are not drawn to scale.

Membrane-aerated biofilm reactors (MABRs)

A new sewage treatment facility is planned for Cambridge (UK) where I live. This would involve moving the current treatment works to a new site. The core design of the plant will operate as a flow-through activated sludge system. But there will be a significant upgrade from the existing plant: the installation of membrane-activated biofilm reactors (MABRs). As with MBRs, the functional units in MABRs are racks of vertical membrane tubes that are immersed in reactor tanks in flowing sewage. But in this case the membrane tubes have a different role, providing a substrate for bacterial growth as a surface biofilm (Figure 19). The centre of each membrane tube is filled with low-pressure pumped air. From here, oxygen diffuses outward through the membrane directly to the bacterial biofilm growing on the outside of the membrane tube. The biofilm bacteria that are closest to the membrane surface will operate aerobically, while the outside layer of biofilm will be low in a low-oxygen environment and will metabolize anaerobically. So, nitrate can be removed through the twin processes of nitrification and denitrification. The rest of the tank is typically a conventional activated sludge plant that further breaks down the organic matter.

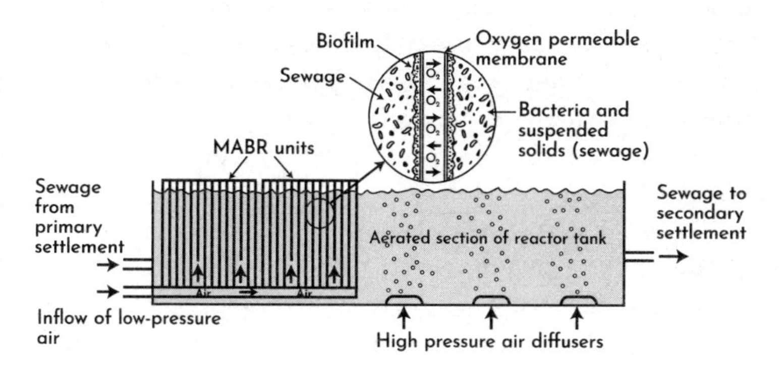

Figure 19. Membrane-aerated biofilm reactor (MABR). Bacterial decomposition of sewage takes place on membrane-attached biofilm, which has both aerobic and anaerobic zones.*

*Bacteria and suspended solids are not drawn to scale.

Trickling filters

Having implied that there are a variety of new methods for wastewater treatment, we now take something of a step backwards. When you travel in the UK by train, one of the features frequently seen on the edge of towns is sewage treatment plants. Often you will see large circular beds, normally above ground level, which have a centrally pivoted, rotating pipe spraying sewage liquor on to the surface of a filter bed that acts as a biofilter – known as a trickling filter (Figure 20). What we have here is a fixed bed of coarse material like granite rocks, slag, coke or artificial media. It is important that the medium is porous, with large air spaces between the packing material that provides a large surface area. The filling material develops a surface biofilm of microorganisms and other organisms, similar to those described for activated sludge. The surfaces must remain constantly wet with a stream of sewage trickling through the bed. Air either gets pumped through the filter or, more typically, diffuses passively through the filter medium, both from the surface and the base. The circular motion of the sprinkler pipe ensures that air can circulate into the filter bed in the intervals after the pipe passes, when that part of filter bed drains of sewage.

Treatment in these trickling filters can be good in terms of breakdown of organic matter, but they are less well suited to nutrient removal – although adaptations are possible. As well as the circular versions described, there are rectangular versions, with the pipework distributing the sewage moving forwards and backwards across the surface of the filter bed.

While such trickling filters are thought of as old technology, they are increasingly being seen as a sustainable wastewater treatment system and so are coming back into favour. They do require quite a lot of land to provide enough treatment capacity, but they are simple to construct and use readily available materials for the filter bed, which does not need replacement. They require little maintenance and typically use little energy to run.

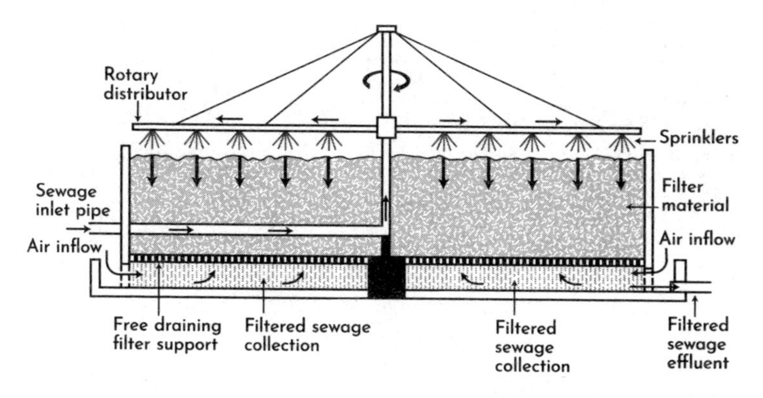

Figure 20. Schematic cross-section of a trickling filter. Sewage percolates through the permeable medium from the top surface, while air flows upward from the base.

Rotating biological contactors (RBCs)

Some sewage treatment plants can be smaller in scale – for example, those that service a small rural population. This is a typical scenario suited to the installation of what are known as rotating biological contactors (RBCs). These have been used since the 1960s, following design and proof of concept in (West) Germany. Designed for biological sewage treatment, they are an alternative to trickling filters and activated sludge. The system consists of a series of large discs (typically 2–4 m/7–13 ft across) mounted on a rotating shaft positioned just above the surface of sewage flowing through a rectangular tank (Figure 21).

The discs dip into the sewage as they rotate, such that about 40 per cent of the disc surface is in the sewage liquor at any one time. The rate of rotation is slow, about two to five revolutions per minute. The surface of the discs is covered with a biofilm of bacteria and other organisms, which are alternately exposed to the air (re-oxygenated) before being dunked in the anaerobic sewage mix. The trick is to ensure that the bacteria are

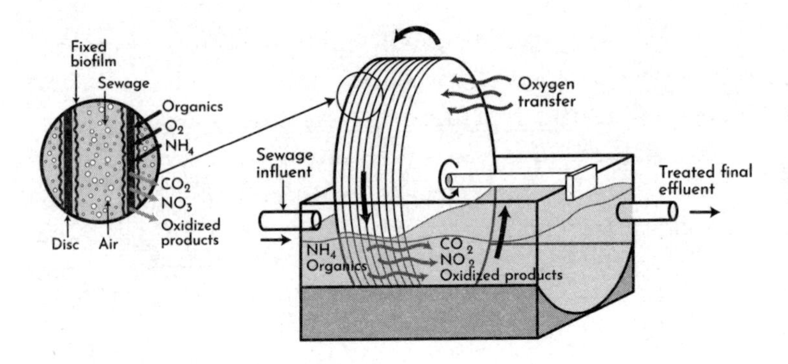

Figure 21. The construction and functioning of a rotating biological contactor (RBC). (Adapted, with permission, from Biorotor S.A.S.)

in the sewage only as long as their oxygen lasts to support the breakdown of organic materials. Typically, there will be two or three or several RBC units linked together to optimize sewage treatment.

RBCs can generate good-quality effluent, albeit lacking the specific function of nitrate and phosphorus removal. They have the added virtue of a relatively small footprint and are typically housed under a canopy roof. Being fully enclosed, they are protected from cold winter conditions.

Sequencing batch reactors (SBRs)

In Chapter 3, about the history of wastewater treatment, we discussed the story of sewage treatment in the UK city of Liverpool. At the tail end of that piece there was reference to the new treatment plant commissioned in the last few years to treat sewage that previously flowed into the River Mersey. Instead of conventional activated sludge tanks, the design here employed sequencing batch reactors. Imagine a series of rather tall, metal or concrete reactor tanks pumped full of a batch of sewage and aerated. In a manner similar to the activated sludge process described earlier, decomposition (oxidation) of organic matter occurs over a period of time, after which the tank is drained. Sludge is drawn off from the base of the tank and doesn't require a separate settlement stage. With a series of tanks, these can be filled, processed and emptied in sequence. So, this is not a continuous flow-through but a batch process. This system has the advantage that tank filling can vary depending on volumes of sewage. Less sewage input from the sewerage system fills fewer tanks.

Constructed wetland (reedbed) sewage treatment

We described activated sludge sewage treatment as managing a microbial ecosystem (microbiome). Now we can move on to a treatment process that is a more familiar biological system. Birdwatchers are used to visiting reedbeds for their ornithological interest. However, in the last forty years, reedbeds and other forms of wetland have been developed as a benign way of contributing to sewage treatment. Although often referred to by the generic term 'constructed wetland', we will focus on the most common type in temperate regions – a constructed wetland of reed plants. This refers to vegetation dominated by the common reed (*Phragmites australis*), an aquatic plant with a worldwide distribution.[9] Initially the emphasis was on finishing off the treatment of sewage by passing it through a reedbed. This process is sometimes referred to as 'polishing', prior to effluent discharge. But it can also be a method of cleaning up diluted storm-water overflow sewage or to clean household grey water. So how do they work – and *do* they work?

---

9 Other species/mixtures of semi-aquatic emergent plants can also be used.

The first point to emphasize is that this is not simply a case of finding a convenient bit of reedbed and adding a sewage outfall pipe at one end. The reedbeds are artificially constructed and engineered for wastewater, providing it with an effective route through the bed so it may exit as 'cleaned' wastewater that meets the appropriate environmental standards for discharge to a water body. Effective wastewater flow through the reedbed and adequate reedbed size are crucial to the proper operation of this method. Wastewater that transitions too quickly (for example, as surface flow rather than through the growth medium) may not be treated adequately.

An outline of a reedbed-type construction is shown diagrammatically in Figure 22. The basic principle is straightforward, with the reedbed acting as a biofilter – in some ways not dissimilar to trickling filters previously described. The bed is filled with a suitable permeable growth medium for the vegetation, such as gravel. The plant growth medium must allow for percolation of wastewater and diffusion of oxygen. If the medium is insufficiently permeable then wastewater will bypass much of the treatment zone and there will be little oxygen around the plant roots. The reedbed shown in Figure 22 is set up for horizontal flow of wastewater – in one end and out at the other. However, other bed designs have vertical flow of wastewater through the reedbed, which is generally more effective in treating wastewater with higher BOD values and nutrient levels. But all this depends on informed design, construction and maintenance of the reedbed.

So, what goes on in the reedbed as wastewater flows through it? The purpose of the treatment is to remove organic matter, nutrients (especially nitrogen and phosphorus) and pathogens. Within the bed there will be a mix of mineral material, organic matter, a microbial community (microbiome) as well as the subterranean parts of the vegetation. These are likely to be mainly roots but may include other structures such as rhizomes, as in the case of reedmace (*Typha latifolia*), which is a plant used in reedbeds.

If it was possible to delve into the reedbed at a microscopic level, it would be clear that there is a great deal of complexity in the arrangement of the living and non-living components. Microorganisms would include large numbers of species. Depending on the oxygen in various parts of the bed, there will be active aerobic and anaerobic bacteria present. You would expect more anaerobes in the deeper parts of the bed (those further from the surface), but the picture is complicated by the presence of plant roots. It is important to remember that the root tissues of plants that grow in wetlands have a requirement for oxygen. Because oxygen diffuses very slowly through a water-saturated soil, wetland plants have specialized tissues in their stems where the cells are loosely packed, leaving air spaces. This type of plant tissue is known as aerenchyma and is a typical adaptation to wetland living, as is the case in reed plants.

These aerenchyma tissues provide a conduit for oxygen to diffuse from the aerial parts of the plant down into the roots. Because some of this oxygen will escape from roots into the surrounding medium, there may be a thin film of aerobic bacteria on the root surface even if it is surrounded by a low-oxygen soil and anaerobic bacteria. The combination of larger roots, fine roots and root hairs and the surrounding soil medium forms an aerobic/anaerobic zone within the soil; this is known as the rhizosphere.

As wastewater passes through the reedbed there will be a simple filtration effect for organic particles and pathogenic bacteria. If the bed is working effectively, loss of some nitrogen is likely through a combination of nitrification and denitrification, as described in Chapter 9. Soluble phosphorus (orthophosphate) may be precipitated by reacting with iron or calcium or aluminium present in the growth medium. Because availability of soluble phosphorus frequently limits growth of plants, it will also be readily taken up by reeds or other plants in the bed. A more recent innovation is the addition of the mineral apatite to the reedbed growth medium to adsorb phosphorus and reduce its concentration in reedbed effluent. Reeds can also take up heavy metals such as copper, cadmium, chromium, zinc and lead and other toxins. The degree of removal will vary with the type of bed, the retention time of wastewater in the bed and other factors.

You might think that the plants are central to this process, but in fact their role is supportive rather than key to the treatment. They do take up some nutrients as they grow, but it is the microbial community that does much of the 'heavy lifting' in terms of mineralization of elements (conversion from organic to inorganic forms) and, in the case of nitrogen, nutrient removal by denitrification.

The importance of constant monitoring and adjustment of treatment-plant performance to achieve good-quality effluent has already been mentioned in relation to activated sludge plants. Most reedbed systems have little scope for such control. But they do have to be carefully designed to provide the capacity for the volume of wastewater and concentration of organic waste (BOD). This needs to take account of the fact that reedbeds are less efficient during cold winter temperatures. Not only are bacteria metabolizing less effectively at lower temperatures, but plant growth has also stopped. During very cold spells the quality of effluent leaving the bed may be compromised. This can be an issue if discharge permits do not allow such variations in effluent quality during the year.

While there are potential limitations of the system as noted, there is also a positive aspect – it can be described as a 'nature-based' solution to wastewater treatment rather than a technological one. It provides wetland habitat that may prove attractive to a range of bird and invertebrate species and potentially enhance biodiversity in the area (Feest et al., 2012).

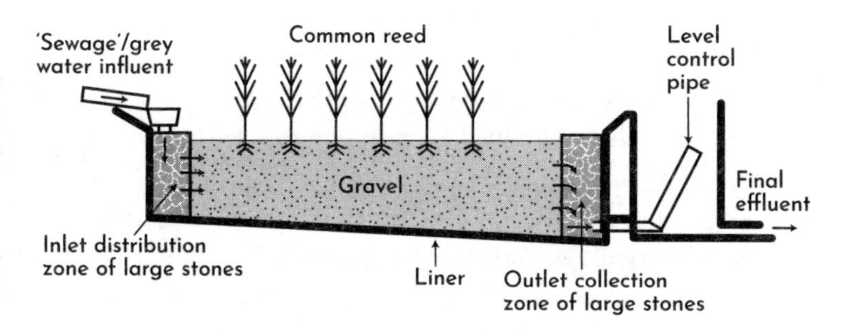

Figure 22. Horizontal-flow constructed reedbed wetland, which is typically used for treating wastewater with a low organic content.

Sewage sludge

Although clean effluent is one outcome of sewage treatment, the other (potentially problematic) outcome is sewage sludge. Sludge is removed during both primary and secondary settlement/sedimentation stages. In primary settlement it will mainly be the heavier organic sewage material that settles out. In secondary settlement there will be further removal of finer organic matter in suspension, as well as live bacteria, dead bacteria and floc. If chemical removal of phosphorus is used at this stage, then this creates additional sludge in the form of precipitated phosphorus. Treating and disposing of sewage sludge is a major cost in wastewater treatment and can be equal to that of sewage treatment.

Although there will be much less solid sewage sludge than the original mass of solids in sewage, the organic material that is harder to decompose is still present. What happens to all this sludge? We had the example of London sewage sludge being dumped in the Thames Estuary. Nowadays sludge is 'digested' within the sewage treatment plant. But before this can be done, the water content of the sludge needs to be reduced. A variety of dewatering methods, such as filtering and centrifugation, can be used. This brings down the water content to leave the sludge much more solid. Sludge treatment needs its own system, quite separate from the main sewage treatment process (see Figure 9). It is typically based on anaerobic digestion – a process of decomposition in closed oxygen-free digestion tanks. The duration of that digestion depends on temperature. At higher sludge temperatures (around 55°C/131°F) there will be 'thermophilic' digestion by heat-tolerant bacteria; the more usual 'mesophilic' digestion with a different suite of bacteria will occur at lower temperatures (25–35°C/77–95°F). Digestion results in the release of carbon as gaseous carbon dioxide (35–40 per cent), mixed with methane (CH_4) gas (60–65 per cent), together known as 'biogas'. The production of methane is due to the activity of methanogenic bacteria-like organisms called the Archaea. Removal of carbon dioxide from biogas leaves 'biomethane', which can be fed into the UK natural gas grid or used as a fuel within the treatment plant. The reduced quantity of digested sludge (about 70 per cent or less of the original volume) is now referred to as less noxious-sounding 'biosolids'. Biosolids can be applied to farmland as a useful organic soil conditioner – which is somewhat reminiscent of the application of 'night soil' to farmland! In the UK and many other countries there is a deficit of organic matter in intensively managed agricultural soils, so farmland application seems a good disposal solution for sludge.

Processed sludge is typically rich in phosphorus and other plant nutrients. Recycling of phosphorus is important because of limited world availability of mined phosphorus for fertilizer production. Currently in the UK, 96 per cent of sewage sludge is disposed of to farmland. This sounds positive, but there is a potential catch.

Although spreading processed sludge on farmland has been done for many years, this practice is becoming more contentious. This is because there are several environmental issues that can be associated with the practice. The obvious issue of disease-causing bacteria (pathogens) is generally not considered to be a problem in treated sludge. On the other hand, heavy metals, other toxins, pharmaceuticals and microplastics are among hundreds of chemicals that may be present in sludge. Recent concerns relate to per- and polyfluoroalkyl substances (PFAS), which have the emotive terms 'forever chemicals' applied to them because they do not break down in the environment. Some of these chemicals could potentially be taken up by crop plants or get washed out (leached) from soils into water courses. Some studies have shown that contaminants from sludge-treated farmland can appear in agricultural produce. In the UK there is a requirement for testing concentrations of elements such as chromium, cadmium, lead, mercury and several others. There are maximum permissible concentrations for application of sludge to farmland. The situation has improved because levels of many potentially toxic elements in sludge have declined as industry has declined in the UK. Nevertheless, application of sludge to land remains a concern, and some countries have banned the practice altogether. In the UK, the Environment Agency is consulting on new rules for sludge disposal, focusing on 'risk-based and outcome-focused regulation' (Environment Agency, 2023).

In principle, tightly controlled recycling of 'safe' sludge on to land is a better alternative than other methods such as landfill. But this approach does require rigorous checks on potentially hazardous batches of sludge. The following is a telling comment made by David Lewis, a former scientist at the US Environmental Protection Agency: 'Spending billions of dollars to remove hazardous chemicals and biological wastes from water, only to spread them on soil everywhere we live, work and play defies common sense' (Perkins, 2019). The avoidance of long-term contamination of soil is as important as the avoidance of contamination of waterbodies.

There are other ways of dealing with sludge. It does have an energy content, so it can be incinerated after reducing the water content. With appropriate pollution safeguards (flue-gas cleaning) it can be a viable alternative to landfill, but is little used in the UK. There is still the issue of potentially contaminated ash disposal to contend with and little public enthusiasm for siting of new incinerators.

The current buzzword in sludge research circles is 'biochar'. This means heating the sludge to several hundred degrees centigrade – a process of pyrolysis – to convert the organic sludge into a charcoal-like material known as biochar. This end-product of pyrolysis is being touted as a valuable soil additive that can help to improve soil texture and crop yields. But, as with applying sewage sludge to farmland, questions remain about the issue of heavy metals that could still be present in some biochar products.

A very basic method of sludge disposal is to pump it into large sludge lagoons, where a combination of drainage and evaporation gradually dries the sludge. This avoids the cost of dewatering but doesn't solve problems of toxic content. It requires lots of land and the drained effluent may contribute to pollution if not treated.

To sum up, the issue of sludge disposal remains difficult. In the UK, if there was any significant restriction on the use for farmland for sludge disposal it would cause huge logistical problems for treatment works. There are no ideal or easy solutions to sludge disposal that are problem-free and cost-effective.

Chapter 11
Sewage Discharge, Effluent Quality and Environmental Impacts

At the end of the sewage treatment process, broadly 'clean' effluent can be released into a receiving water body – often a river, but it may be a lake or the sea. Most of the time these discharges will be of fully treated sewage, but sometimes sewage treatment plants with combined sewer inflows may exceed the capacity of the sewerage and treatment system and release diluted sewage. We start the next section with a brief review of the issue of untreated sewage discharge, followed by a consideration of the ecological impact of treated sewage effluent on receiving water bodies.

Combined sewer overflows (CSOs)

We noted in Chapter 8 that release of raw sewage into a water body can have very significant biological and ecological impacts at the point of discharge and, for a river, some distance downstream as the biology of the river gradually recovers. As raw sewage travels directly to a treatment plant, sewage discharge from the treatment plant should happen only if there is a major technical malfunction. However, there are unplanned releases of untreated sewage into rivers, often due to rainwater surges in combined sewer systems. Combined sewerage systems amount to approximately 100,000 kilometres/62,000 miles of sewer pipes. Treatment plants may have storm tanks to buffer fluctuations of sewage volume caused by rainfall, but these are often insufficient to cope with large volumes of rain. When this happens, combined raw sewage and rainwater overflow into an adjacent water body, typically a river – through combined sewage overflows (CSOs), of which there are about 15,000 located in England alone. At that point there is no alternative if the sewerage system is not to 'explode' because of excessive volumes of sewage and cause potentially devastating flooding by sewage in urban areas. This issue causes a great deal of public anxiety because the notion of sewage flowing into a river seems like ecological vandalism. And there are questions to be asked over claims of excessive release of sewage overflow by water companies when there is little or no rainfall. While CSOs are far from ideal, they can only be changed by a massive expenditure on additional storm-tank capacity, as has been done in London and Paris (Chapter 5).

Combined sewerage systems create a variety of problems. The inflow of any sewage to any aquatic system is bound to have deleterious effects, but if diluted with rainwater the impact will depend on the extent of dilution and the 'sensitivity' of the receiving water body. Some waters are more pristine than others and with a more specialized and vulnerable fauna and flora.

But combined sewerage systems can also cause difficulties in sewage treatment plants. We have already mentioned the importance of the food to microorganisms (F/M) ratio in keeping a steady and active bacterial population (Chapter 9). A sudden influx of rainwater will potentially play havoc in the concentration of both food and microorganisms in, for example, an activated sludge reactor.

The government agency with responsibility for sewage in the UK is the Department of Environment, Food and Rural Affairs (Defra). They have recently published a plan that requires improvements in sewage release from combined systems (Defra, 2022). This does not advocate the wholesale redesign of sewerage systems and few prescriptive solutions are listed. But it does set out requirements for water companies to do better – with a timescale running up to 2050. The document sets four main targets:

1. 'Water companies will only be permitted to discharge from storm overflow where they can demonstrate there is no local adverse ecological impact.'

   Comment: the word 'no' in this statement is likely to prove challenging! Is this even possible? It is surprising that 'no' has not been replaced by the ambiguous term 'significant' used in Target 2.
2. 'Water companies must significantly reduce harmful pathogens from storm overflows discharging into and near designated bathing waters, by either: applying disinfection; or reducing the frequency of discharges to meet Environment Agency spill standards by 2035.'
3. 'Storm overflows will not be permitted to discharge above an average of 10 rainfall events per year by 2050.'
4. 'Water companies will be required to ensure all storm overflows have screening controls.'

   Comment: This would be to filter out visible 'solids' before discharge but not other components of sewage and may be difficult to implement at all overflows.

Various environmental pressure groups remain unimpressed by the proposed speed and scope of change.

Biological impacts of sewage treatment effluent

In Chapter 9 we discussed the role of sewage treatment in reducing the nutrient load of discharged sewage effluent. Although nutrient removal is applied in many treatment plants, there is generally no requirement to do this for smaller plants. It is also technically not possible to ensure 100 per cent removal of all nitrogen and phosphorus in final sewage effluent. So, some nitrogen and phosphorus will still enter receiving waters – coupled with additional inputs from diffuse agricultural sources that are harder to control. Indeed, agricultural sources of phosphorus have taken over from sewage treatment works as the main source of phosphorus in natural water bodies. On the positive side, levels of phosphorus releases from these and other sources are declining. Whether they continue to do so will depend on the regulatory environment, especially from diffuse pollution sources such as farmland managed for livestock and crops.

Why should the presence of these two nutrients be an issue in sewage-effluent release? After all, both these nutrients are desirable components of fertilizers and enhance crop production – hence the use of nitrogen and phosphorus fertilizers in crop farming. But this is also a clue to what these nutrients do to a river or lake that may be naturally nutrient poor. Where this is the case, the ecological community present in these rivers or lakes – both plant and animal – are adapted to relying on tight recycling of available nutrients. You could say that the ecosystem is stable and 'in balance'. The addition of nutrients – nitrogen and especially phosphorus – upsets this balance. Because phosphorus is insoluble in many of its chemical forms, available (that is, soluble) phosphorus is often a key limiting factor for productivity in natural ecosystems. The addition of extra nutrients to water from sewage or sewage effluent (nutrient enrichment) is known scientifically as eutrophication (Chapter 8). It can have very deleterious effects on the aquatic environment. The impact begins with a growth response from aquatic plants and other organisms. Most visible are larger aquatic macrophyte plants (sometimes referred to, rather inappropriately, as water weeds), but equally importantly are algae and bacteria. There is typically a rapid upsurge in growth of all three taxa.

In rivers, increased plant growth can impede navigation and incurs the cost of weed-cutting to keep rivers open. If the plants are left to grow, they will typically die back in autumn and all that extra rotting organic matter may impose an extra BOD penalty on the river. It is a more subtle version of what happens when sewage enters the river, albeit rather less drastic and spread over a longer period of time. Some aquatic detritivore invertebrates may even enjoy an unexpected feast of dead plant material!

It is algae and/or blue-green bacteria (cyanobacteria – often incorrectly called blue-green algae) that can turn a water body into a pea-green 'soup' – a phenomenon that is particularly visible in lakes but may also be seen in slow-flowing rivers and coastal waters. These are referred to as algal blooms. One consequence of this is that aquatic macrophytes (water weeds) are cut off from light and may start to die off. And as they die, so will the associated animal life. But as far as the public are concerned, a bigger worry is the release of toxins into the water by algae or cyanotoxins by blue-green bacteria. There are many different types of such toxin and different modes of toxicity – including some that are neurotoxic and others that are carcinogens. They are harmful to people and their pets, but also to aquatic invertebrates and vertebrates such as fish. The presence of algal bloom toxins is a major reason for bathing waters and other recreational water bodies being periodically closed off.

During the summer months, the release of bacterial and algal toxins can have drastic effects on rivers. They can cause mass mortality of fish, as happened catastrophically in the Odra River on the border of Poland and Germany. Here, in the summer of 2022, a bloom by golden algae (*Prymnesium parvum*) caused the release of a toxin that destroys blood cells (haemolytic) and that led to the death of at least 249 tonnes (249,000 kg) of fish between late July and mid-September. It was described as an ecological disaster (Sobieraj and Metelski, 2023). It is not unusual for algal blooms to be reported in the UK, albeit on a smaller scale. A high-profile example is Lake Windermere in the English Lake District. Here, algal blooms of cyanobacteria have occurred with increasing frequency in summer. The lake receives nutrient inputs from sewage treatment plant overflows but also from agriculture and lakeside septic tanks.

At the end of the summer season, blue-green bacteria and algae cause further problems. Having absorbed excess nutrients from the water, the cells start to die off and generate a demand for oxygen as they decay. This can deplete still lake water and slow-flowing rivers of oxygen and cause additional mortality of vertebrate and invertebrate animal life. However, deoxygenation doesn't affect all species equally. As discussed in Chapter 7, invertebrates vary in their response to organic river pollution and dissolved oxygen. Some – such as the non-biting midge larvae (Chironomidae) – are more tolerant of low oxygen levels in water. Among these midge larvae are the bloodworms, whose bright-red colour comes from the pigment haemoglobin (also present in human blood cells), which has a high affinity for oxygen. This facilitates the survival of these larvae and other adapted invertebrates in waters suffering from the aftermath of decay of plant material and bacterial blooms. Nevertheless, the lack of oxygen in the water can drastically deplete and change the aquatic invertebrate community structure.

To avoid the problems of excessive plant growth and blooms, a standard for the critical concentration of phosphorus in water that can cause these kinds of problems is needed. One difficulty with measuring phosphorus concentrations is the difference between total phosphorus and what is referred to as reactive phosphorus. Only the latter is biologically active. There is no single limit for release of reactive phosphorus in lakes and rivers – although 1 mg/l (1 mg l^{-1}) is a typical value. The limits depend on the nature of the water body and its sensitivity to eutrophication. But the regular occurrence of algal blooms in a variety of both still and slow-flowing waters suggests that aquatic phosphorus concentrations continue to be a problem. Some are attributable to sewage treatment, but there are also other sources such as the fertilizers used in farming.

While the major organic components of sewage can be dealt with effectively through conventional wastewater treatment, the same cannot be said for novel substances (so-called xenobiotics) such as pharmaceuticals and pesticides or legacy chemicals such as chlorinated biphenyls (PCBs) that are persistent and cannot readily be eliminated from effluent water. Such chemicals, even in low concentrations, may compromise the recovery of rivers following improvements in other aspects of sewage treatment (Windsor et al., 2019). In the longer term it may be possible to eliminate some of these by treating effluent water with ozone in wastewater treatment plants. The European Union has started to consider measures to reduce water pollution caused by pharmaceuticals and cosmetics, but the government in England and Wales is not currently inclined to follow that lead (Horton and Laville, 2024).

In our earlier discussions of wastewater treatment technology, we made big play of the improvements that are now possible in the quality of effluent leaving sewage treatment plants. At the same time, there has been rising public concern in the UK about water quality, especially in UK river systems. So, is there any evidence that the additional investment in treatment plants is having positive effects on water quality? Or is the impact of such improvements being swamped by adverse environmental impacts, such as diffuse runoff pollution from agricultural land? A recently published study used macroinvertebrate BMWP scores and related data/indices to assess changes in invertebrate biodiversity in England over a thirty-year period of monitoring (Qu et al., 2023). Despite justified public concern, the positive news is that there has been a significant overall improvement in macroinvertebrate biodiversity over this period – evidence that environmental degradation need not be a one-way street. This analysis was based on family-level (rather than species-level) invertebrate surveys. In that sense it may not present a full picture of what is happening to the more sensitive invertebrate species. Nevertheless, the study does suggest that investment in improved wastewater treatment does generate benefits in water quality and biodiversity. However, the recent (2024) report by the Rivers Trust (*The State of Our Rivers Report*) noted that only 15 per cent of rivers in England had good or better ecological status. There is still more work to be done.

Chapter 12
The Last Word . . .

This book has primarily been concerned with the progressive development of wastewater treatment methods over time and has given an explanation of the technology involved. Some of the details may have been a little taxing to follow, but hopefully they have shed light on wastewater treatment as a challenging process. This especially true when living organisms represent the core of the processing system. Unlike a factory, where there are fairly predictable inputs and outputs, a wastewater treatment plant has to cope with the vagaries of sewage flows related to time of day and time of year and of rather unpredictable weather at the local level. The profile of sewage entering each treatment works is different depending on a host of factors such as geographical location – for example, town versus rural or industrial versus agricultural. At its best, wastewater treatment can increasingly produce good-quality effluent that can be released into an aquatic environment in line with discharge limits. That isn't to underplay concern about releases of sewage into rivers, lakes and the sea, or about the host of novel chemicals (xenobiotics) and micro- and nanoplastics that are now part of modern life and pass through most current wastewater plants. Xenobiotics in particular can be released into aquatic environments with limited knowledge of their biological effects and have yet to be tackled by the wastewater industry.

Having said all that, there are still reasons to appreciate the immense contribution that wastewater treatment has made to society. It is our noxious waste that we are happy to send down the sewer pipes for someone else to sort out. It is hard to imagine the filth, squalor and disease that lack of sanitation and sewage treatment generated in urban environments throughout history. It is still evident in some parts of the world where a proper sanitation system would be a huge advance for large numbers of people in coping with urban living. In 2007, over 11,000 readers of the *British Medical Journal* voted the introduction of clean water and sewage disposal (the 'sanitary revolution') the most important medical milestone since 1840. It was even (just) ahead of the discovery of antibiotics!

In the UK, the role of water companies and the contamination of rivers and seas has become a very contentious issue in recent years. This problem isn't unique to the UK, but in a crowded island the rivers in England especially have major water-quality issues. There is much to do to improve the situation. Although this book was not about apportioning blame for the challenges facing wastewater treatment in the UK, it does allude to the financial cost and the technical difficulties of ensuring that water bodies receive only 'clean' treated effluent. There is a lot of public resentment about the way that water services companies are perceived to be 'getting away' with polluting the environment. Public opinion conflates issues of profit-taking, payment of shareholder dividends, lack of investment and seemingly flouting of sewage release limits to explain this poor water quality in seas and rivers. Perhaps the final comment to make on this issue is that there is a collective responsibility for improving the situation. Ofwat, the UK government, a properly funded Environment Agency as well as rigorously regulated and scrutinized water services companies (private or public) must all contribute. And, of course, the likelihood is that the public will need to accept that they need to pay more or be taxed more for cleaner rivers and seas. That is the bottom line.

Glossary

Activated sludge: Aerated organic sewage material being broken down by aerobic (oxygen-requiring) bacteria.

Aerobic respiration: A process of biochemical breakdown of organic molecules in cells that requires oxygen to release energy that then powers the metabolism of the cell.

Ammonification: The bacterial decomposition of complex nitrogen-containing organic molecules to release ammonia, a simple inorganic molecule that can exist in gaseous form or be dissolved in water.

Anaerobic respiration: A process of partial biochemical breakdown of organic molecules in cells that doesn't require oxygen to release energy that powers the metabolism of the cell. The energy yield is lower than aerobic respiration.

Anoxic: A gaseous or aqueous environment where oxygen is absent.

Assimilation: The absorption of nutrients (for example, nitrogen-containing molecules or ions) by organisms and their incorporation in new organic molecules in cells and tissues.

Biochemical oxygen demand (BOD): A measure of the amount of dissolved oxygen required by aerobic bacteria to break down (oxidize) the organic material in a specified volume of sewage, at a specified temperature (20°C/68°F) and length of time (for example, five days). In sewage there will 'difficult' organic components that do not break down during the period of measurement.

Biofilm: A coating of bacteria together with a secreted matrix (of extracellular polymeric substances) and other microscopic organisms that grows collectively as a film on solid surfaces.

Chemical oxygen demand (COD): A measure of the amount of dissolved oxygen required by strong oxidizing chemical agents to break down (oxidize) the organic material in a specified volume of sewage, at a specified temperature (150°C/302°F) and length of time (2 hours). Because COD uses strong oxidizing agents, it breaks down more of the organic material than the BOD method. So, COD measurements will show a higher requirement for oxygen than BOD.

Colloid: Molecules or particles in an aqueous environment which have properties that mean they are in neither solution nor suspension.

Combined sewer overflow (CSO): The 'safety valve' for a sewerage system when the volume of sewage exceeds storage and treatment capacity. Sewage is typically released into a water body (river, lake or sea).

Constructed wetlands/reedbeds: A constructed wetland can be used for treating sewage with a low concentration of organic matter, such as might come from a combined sewer outflow or 'grey water' from a domestic source.

Cyanobacteria: These were previously referred to as blue-green algae, but are correctly referred to as blue-green bacteria. The prefix 'cyan' refers to their colour – which occurs because of photosynthetic pigments in the cells. Some species produce toxins (poisons) known as cyanotoxins.

Cyanotoxins: Toxins released into the water by cyanobacteria (blue-green bacteria).

Decomposers: Organisms that break down dead or decaying organisms or organic material originating from living organisms.

Denitrification: Bacterial conversion (by anaerobic respiration) of nitrate and nitrite ions to gaseous nitrogen.

Eutrophication: Excessive nutrient enrichment of natural water bodies – typically associated with human activities such as agricultural runoff, raw sewage or treated wastewater pollution.

Invertebrates: Organisms without a backbone – or any bones, for that matter. Covers many different groups, such as worms, insects, crustaceans and nematodes as well as single-celled protozoa (Protoctista).

Ions: An atom or group of atoms carrying a positive or negative charge. For example, an ammonia molecule in solution splits (dissociates) into NH_4^+ and OH^- ions. Another ion is nitrate NO_3^-.

Metabolism: All the basic biochemical processes that function continuously in organisms in cells or tissues to sustain life.

Microplastics: Small polluting plastic particles <5 mm (but larger than nanoplastics).

Mineralization: The conversion of a nutrient element by decomposition from an organic form to a soluble inorganic form that makes it available for uptake by plants and bacteria.

Nanoplastics: Tiny polluting plastic particles <1 µm (1 mm = 1,000 µm).

Nitrification: The conversion of inorganic ammonia to nitrite and nitrate ions by bacteria.

Ofwat: The official government body for England and Wales that regulates prices, investment and operational requirements of water services companies.

Oxidation: In the context of sewage treatment, this is the conversion of molecules through the addition of oxygen to the products – but the process has a more technical meaning related to the loss of electrons.

Pathogen: A disease-causing organism.

Pollution: The introduction of harmful materials into the environment. Typically relates to the activities of humans but can be natural.

Protoctists (protozoa): The terms Protoctista and protozoa refer to single-celled eukaryote organisms (cells with a nucleus). Both terms are considered redundant by taxonomists because they do not form a natural evolutionary grouping, but Protoctista is used in preference to protozoa. These organisms have now been split into several different supergroups.

Sewage: Wastewater produced by a community of people. It is typically dominated by organic waste from urine and faeces, but can also include industrial and agricultural waste.

Sewerage: The system of pipes and pumps that carries sewage from a source (such as a domestic house) to the treatment works.

Sludge (sewage): Heavier organic materials entering wastewater-treatment plants. These materials settle out in primary sedimentation (primary sludge) or undecomposed organic and bacterial material that settles out in secondary sedimentation after activated sludge treatment (secondary sludge). This secondary sludge contains greater quantities of bacterial material that can be difficult to decompose.

Taxon: A unit of classification of animal, plant or microbial organisms. Taxa are organized in hierarchical fashion, starting with large groupings with many species. For example, phylum is very large; class and then order are smaller; and there are many fewer species at the level of family and fewer still at the level of genus.

Toxins: Poisons of one kind or another that may be present in sewage or released by blue-green bacteria in natural waters.

Water services: Companies (private or state-owned) that provide drinking water and/or sewerage services.

Xenobiotics: Chemical substances that are not naturally produced or expected to be found within an organism or in the environment.

Selected References

Ardern, J. E. and Lockett, W. T. (1914) 'Experiments on the oxidation of sewage without the aid of filters', *Journal of the Society of Chemical Industries*, 33, pp. 523–539. (Note: Two further papers extending their studies were published in 1914 and 1915.)

Bhat, S. U. and Qayoom, U. (2022) 'Implications of sewage discharge on freshwater ecosystems', in Zhary, T. (ed.) *Sewage – recent advances, new perspectives and applications*. IntechOpen. Available at: http://dx.doi.org/10.5772/intechopen.100770.

Collinson, A. (2019) 'How Bazalgette built London's first super sewer', *Museum of London*. Available at: www.museumoflondon.org.uk/discover/how-bazalgette-built-londons-first-super-sewer (accessed: 24 February 2024).

Defra (2022) 'Storm overflows discharge reduction plan', *Gov.UK*. Available at: https://www.gov.uk/government/publications/storm-overflows-discharge-reduction-plan (accessed: 29 February 2024).

Edmonds-Brown, V. (2022) 'How did the Thames become one of the world's cleanest city rivers?' *The Independent*. Available at: https://www.independent.co.uk/news/science/how-thames-clean-river-citiy-b2064862.html (accessed: 24 April 2024).

Environment Agency (2023) Environment Agency strategy for safe and sustainable sludge use. *Gov.UK.* Available at: https://www.gov.uk/government/publications/environment-agency-strategy-for-safe-and-sustainable-sludge-use (accessed: 24 April 2024).

Feest, A., Merrill, I. and Aukett, P. (2012) 'Does botanical diversity in sewage treatment reed-bed sites enhance invertebrate biodiversity?', *International Journal of Ecology*, Article ID 324295, 9 pages.

Guardian Editorial (2023) 'The Guardian view on English water companies: a badly broken system', *The Guardian.* Available at: https://www.theguardian.com/environment/commentisfree/2023/may/21/the-guardian-view-on-englands-water-companies-a-badly-broken-system (accessed: 24 April 2024).

Halliday, S. (1999) *The Great Stink of London.* Stroud: Sutton Publishing.

Hitchman, J. (ed.) (1866) *The Sewage of Towns.* London: Simpkin, Marshall & Co.

Horton, H. and Laville, S. (2024) 'England won't adopt EU river pollution rules for pharma and cosmetics firms', *The Guardian.* Available at: https://www.theguardian.com/environment/2024/mar/22/england-wont-adopt-eu-river-pollution-rules-for-pharma-and-cosmetics-firms (accessed: 22 March 2024).

Hynes, H. B. N. (1960) *The Biology of Polluted Waters.* Liverpool: Liverpool University Press.

Kawa, N. C. et al. (2019) 'Night soil: Origins, discontinuities, and opportunities for bridging the metabolic rift', *Ethnobiology Letters*, 10, pp. 40–49.

Laville, S. (2024) 'How could England's water system be fixed?', *The Guardian*. Available at: https://www.theguardian.com/business/2024/jan/10/how-could-englands-water-system-be-fixed (accessed: 10 January 2024).

Leach, A. et al. (2023) 'Down the drain: How billions of pounds are sucked out of England's water system', *The Guardian*. Available at: https://www.theguardian.com/environment/ng-interactive/2022/dec/01/down-the-drain-how-billions-of-pounds-are-sucked-out-of-englands-water-system (accessed: 28 June 2023).

Macadam, C. (2021) 'Co-variance of invertebrate biotic indices indicated by an analysis of Environment Agency monitoring data'. Unpublished. (Data discussed in Murray-Bligh and Griffiths, 2022.)

Murray-Bligh, J. and Griffiths, M. (2022). *Freshwater Biology and Ecology Handbook: Practitioners' Guide to Improving and Protecting River Health; Focus on Invertebrate Monitoring and Assessment*. Marlow: Foundation for Water Research; Ulverston: Freshwater Biological Association.

Paisley, M. F. et al. (2014) 'Revision of the biological monitoring working party (BMWP) score system: Derivation of present-only and abundance-related scores from field data', *River Research and Applications*, 30(7), pp. 887–904.

Perkins, T. (2019) 'Biosolids: Mix human waste with toxic chemicals, then spread it on crops', *The Guardian*. Available at: https://www.theguardian.com/environment/2019/oct/05/biosolids-toxic-chemicals-pollution (accessed: 24 April 2024).

Qu, Y. et al. (2023) 'Significant improvement in freshwater invertebrate biodiversity in all types of English rivers over the past 30 years', *Science of the Total Environment*, 905(2023) 167144.

Richardson, M. and Soloviev, M. (2021) 'The Thames: Arresting ecosystem decline and building back better', *Sustainability*, 2021,13,6045.

The Rivers Trust (2024) *The State of Our Rivers Report.* Available at: https://theriverstrust.org/rivers-report-2024 (accessed: 29 February 2024).

Salvidge, R. and Hosea, L. (2023) 'Revealed: Scale of "forever chemical" pollution across UK and Europe', *The Guardian.* Available at: https://www.theguardian.com/environment/2023/feb/23/revealed-scale-of-forever-chemical-pollution-across-uk-and-europe (accessed: 24 April 2024).

Sobieraj, J. and Metelski, D. (2023) 'Insights into toxic *Prymnesium parvum* blooms as a cause of the ecological disaster in the Odra River', *Toxins*, 15(6), 403.

Tudor, S. (2022) 'In Focus: Sewage pollution in English waters', *House of Lords Library.* Available at: https://lordslibrary.parliament.uk/sewage-pollution-in-englands-waters/ (accessed: 24 April 2024).

WHO/UNICEF (2015) 'Progress on sanitation and drinking water: 2015 update and MDG assessment'. Geneva: WHO Press.

Windsor, F. M. et al. (2019) 'Persistent contaminants as potential constraints on the recovery of urban river food webs from gross pollution', *Water Research*, 163(2019) 114858.

Acknowledgements

Any opinions or views expressed in this book relating to water services are my own and do not represent the views of UK water services companies. I am grateful for the technical discussions I have had with Dr Ralitza Nikolova-Kuscu, Senior Associate Process Engineer at the consultancy company Mott Macdonald; Martin Jolly, Engineering Technical Consultant at Yorkshire Water; and Rowan Luck, Treatment Process Engineering Team Leader at Severn Trent Water. These discussions have all related to clarification of technical details relating to wastewater treatment and have been very helpful for that aspect of the book. Any errors of interpretation of technical details remain my own.

The process of writing and producing the book has required collective input from my wife Christine, who read the first draft, professional editing by Monica Hope, figures and illustrations by Sarah Pyke and John Gilkes, overall coordination and management of the book production process by Anna Sanderson. All are thanked for their expert contributions.

Index